Springer Theses

Recognizing Outstanding Ph.D. Research

Aims and Scope

The series "Springer Theses" brings together a selection of the very best Ph.D. theses from around the world and across the physical sciences. Nominated and endorsed by two recognized specialists, each published volume has been selected for its scientific excellence and the high impact of its contents for the pertinent field of research. For greater accessibility to non-specialists, the published versions include an extended introduction, as well as a foreword by the student's supervisor explaining the special relevance of the work for the field. As a whole, the series will provide a valuable resource both for newcomers to the research fields described, and for other scientists seeking detailed background information on special questions. Finally, it provides an accredited documentation of the valuable contributions made by today's younger generation of scientists.

Theses may be nominated for publication in this series by heads of department at internationally leading universities or institutes and should fulfill all of the following criteria

- They must be written in good English.
- The topic should fall within the confines of Chemistry, Physics, Earth Sciences, Engineering and related interdisciplinary fields such as Materials, Nanoscience, Chemical Engineering, Complex Systems and Biophysics.
- The work reported in the thesis must represent a significant scientific advance.
- If the thesis includes previously published material, permission to reproduce this must be gained from the respective copyright holder (a maximum 30% of the thesis should be a verbatim reproduction from the author's previous publications).
- They must have been examined and passed during the 12 months prior to nomination.
- Each thesis should include a foreword by the supervisor outlining the significance of its content.
- The theses should have a clearly defined structure including an introduction accessible to new PhD students and scientists not expert in the relevant field.

Indexed by zbMATH.

More information about this series at https://link.springer.com/bookseries/8790

Mourad Elsobky

Ultra-Thin Sensors and Data Conversion Techniques for Hybrid System-in-Foil

Doctoral Thesis accepted by
University of Stuttgart, Germany

Springer

Author
Dr. Mourad Elsobky
Institute of Nano and Microelectronic Systems (INES)
University of Stuttgart
Stuttgart, Germany

Supervisor
Prof. Dr.-Ing. Joachim N. Burghartz
Institute of Nano and Microelectronic Systems (INES)
University of Stuttgart
Stuttgart, Germany

ISSN 2190-5053 ISSN 2190-5061 (electronic)
Springer Theses
ISBN 978-3-030-97725-2 ISBN 978-3-030-97726-9 (eBook)
https://doi.org/10.1007/978-3-030-97726-9

This Springer imprint is published by the registered company Springer Nature Switzerland AG
The registered company address is: Gewerbestrasse 11, 6330 Cham, Switzerland

Dedicated to my beautiful wife, Soheila,
beloved parents and kind sister, Amira

Supervisor's Foreword

Flexible electronics features a paradigm shift by moving away from rigid assembly of electronic components to a fully flexible, form adaptive, large-area and even bendable form factor. While the large-area aspect requires implementation of electronic devices and circuits different from the established compact microelectronics integration, it remains highly desirable keep high-performance chip electronics involved for providing an overall optimum system architecture and performance.

This leads to Hybrid Systems-in-Foil (HySiF), aiming at a combination of ultrathin chips, thin-film devices and sensors, flexible and bendable interconnects, energy storage and possibly harvesting, as well as wireless communication interfaces, on a flexible substrate which also serves as the final package.

The thesis by Mourad Elsobky describes several advancements in engineering HySiF systems. This includes an innovative approach in implementing sensory functions off-chip when large area is required versus on-chip when sensors can be compact yet need to be of high quality. The off-chip approach is demonstrated on a new type of humidity sensor. Also, energy-efficient HySiF design is demonstrated through a low-power chip design base on a new scalable architecture of an analog-to-digital converter (ADC).

Stuttgart, Germany
March 2022

Prof. Dr.-Ing. Joachim N. Burghartz

Parts of this thesis have been published in the following documents:

Books and Journal Articles

1. U. Passlack, **M. Elsobky**, Y. Mahsereci, C. Scherjon, C. Harendt, J. N. Burghartz:
   "A Flexible Chip-Film Patch and a Flexible Strain Gauge Sensor Suitable for a Hybrid Systems-in-Foil Integration"
   IEEE Sensors Journal, Early Access, 2021
2. **M. Elsobky**, A. Mohamed, T. Deuble, J. Anders, J. N. Burghartz:
   "A 12-to-15 b 100-to-25 kS/s Resolution Reconfigurable, Power Scalable Incremental ADC"
   IEEE Sensors Letters, vol. 5, no.2, pp. 1–4, Feb., 2021
3. **M. Elsobky**, T. Deuble, S. Ferwana, B. Albrecht, C. Harendt, A. Ottaviani, M. Alomari, J. N. Burghartz:
   "Characterization of On-Foil Sensors and Ultra-Thin Chips for HySiF Integration"
   IEEE Sensors Journal, vol. 20, no.14, pp. 7595–7604, July, 2020
4. J. N. Burghartz, G. Alavi, B. Albrecht, T. Deuble, **M. Elsobky**,*et al.*:
   "Hybrid Systems-in-Foil—Combining the Merits of Thin Chips and of Large-Area Electronics"
   IEEE Journal of the Electron Devices Society, vol. 7, no. 6, pp. 776–783, January, 2019
5. A. Eldebiky, **M. Elsobky**, H. Richter, J. N. Burghartz:
   "Humidity and Temperature Sensor System Demonstrator with NFC Tag for HySiF Applications"
   Advances in Radio Science, Copernicus Publications, vol. 16, pp. 109–116, September, 2018
6. **M. Elsobky**, Y. Mahsereci, Z. Yu, H. Richter, J. N. Burghartz, J. Keck, H. Klauk, U. Zschieschang:
   "Ultra-thin smart electronic skin based on hybrid system-in-foil concept combining three flexible electronics technologies"
   Electronics Letters, vol. 54, no. 6, pp. 338–340, March, 2018
7. **M. Elsobky**, Y. Mahsereci, J. Keck, H. Richter, J. N. Burghartz:
   "Design of a CMOS readout circuit on ultra-thin flexible silicon chip for printed strain gauges"
   Advances in Radio Science, Copernicus Publications, vol. 15, pp. 123–130, September, 2017

8. **M. Elsobky**, M. Elattar, G. Alavi, F. Letzkus, H. Richter, U. Zschieschang, M. Strecker, H. Klauk, J. N. Burghartz:
"A digital library for a flexible low-voltage organic thin-film transistor technology"
Organic Electronics, Elsevier, vol. 50, pp. 491–498, November, 2017
9. **M. Elsobky**, J. N. Burghartz:
"Chip-Flex Technology: Hybrid Systems-in-Foil"
Elements in Flexible and Large-Area Electronics, Cambridge University Press, Under Review, 2020

Conference Papers

1. U. Passlack, **M. Elsobky**, A. Mueller, C. Scherjon, C. Harendt, J. N. Burghartz:
"Chip-Film Patch Sensor System with Integrated Read-out ASIC for Biomedical Applications"
IEEE International Conference on Flexible and Printable Sensors and Systems (FLEPS), 2020
2. B. Albrecht, **M. Elsobky**, T. Deuble, C. Harendt, J. N. Burghartz:
"HySiF—hybride Systeme in Folie"
Fachsymposium Hybrid Senor, Karlsruhe, Germany, November, 2019
3. B. Albrecht, **M. Elsobky**, M. Moser, S. Ferwana, C. Harendt, J. N. Burghartz:
"Chip-Film Patch—Packaging Technologie zum Einbetten ungehäuster, funktionaler Siliziumchips in extrem dünne Folien"
MikroSystemTechnik Kongress 2019, Munich, Germany, October, 2019
4. **M. Elsobky**, A. Ottaviani, M. Alomari, Z. Yu, T. Deuble, J. N. Burghartz:
"Characterization of Thin-Film Temperature Sensors and Ultra-Thin Chips for HySiF Integration"
IEEE International Conference on Flexible and Printable Sensors and Systems (FLEPS), Glasgow, UK, July, 2019
5. B. Albrecht, U. Passlack, **M. Elsobky**, C. Harendt, J. N. Burghartz:
"Flexibles Package für die Integration von ultra-dünnen Sensoren und Auslesechips"
microTEC Südwest Clusterkonferenz, Freiburg, Germany, March, 2019
6. **M. Elsobky**, G. Alavi, B. Albrecht, T. Deuble, C. Harendt, H. Richter, Z. Yu, J. N. Burghartz,:
"Ultra-Thin Sensor Systems Integrating Silicon Chips with on-Foil Passive and Active Components"
EUROSENSORS 2018, Proceedings-MDPI, Graz, Austria, December, 2018

7. B. Albrecht, G. Alavi, **M. Elsobky**, S. Ferwana, U. Passlack, C. Harendt, J. N. Burghartz:
"Multi-Chip patch in low stress polymer foils based on adaptive layout for flexible sensor systems"
Electronics System Integration Technology Conference (ESTC), Dresden, Germany, September, 2018
8. J. N. Burghartz, G. Alavi, B. Albrecht, T. Deuble, **M. Elsobky**, et al.:
"Hybrid Systems-in-Foil Combining Thin Chips with Large-Area Electronics"
IEEE International Flexible Electronics Technology Conference (IFETC), Otawa, Canada, August 8, 2018
9. **M. Elsobky** [invited talk]:
"Ultra-thin Sensor Systems using Chip-Film Patch Technology"
Day of Information Logistics, Hochschule für Technik, Stuttgart, Germany, April, 2018
10. **M. Elsobky**, B. Albrecht, T. Deuble, C. Harendt, J. N. Burghartz, P. Ganter, K. Szendrei, B. Lotsch:
"Ultra-thin Relative Humidity Sensors for Hybrid System-in-Foil Applications"
microTEC Südwest Clusterkonferenz, Freiburg, Germany, April, 2018
11. **M. Elsobky**, B. Albrecht, H. Richter, J. N. Burghartz, P. Ganter, K. Szendrei, B. Lotsch:
"Ultra-thin Relative Humidity Sensors for Hybrid System-in-Foil Applications"
IEEE SENSORS 2017, Glasgow, Scotland, UK, October, 2017
12. **M. Elsobky**, B. Albrecht, T. Deuble, C. Harendt, J. N. Burghartz, P. Ganter, K. Szendrei, B. Lotsch:
"Relative Humidity Sensors for System-in-Foil Applications"
MikroSystemTechnik Kongress 2017, Munich, Germany, October, 2017
13. A. Eldebiky, **M. Elsobky**, H. Richter, J. N. Burghartz:
"Humidity and Temperature Sensor System Demonstrator with NFC Tag for HySiF Applications"
Kleinheubacher Tagung, Miltenberg, Germany, September, 2017
14. **M. Elsobky**, Y. Mahsereci, J. Keck, H. Richter, J. N. Burghartz:
"Design of a CMOS Readout Circuit on Ultra-thin Flexible Silicon Chip for Printed Strain Gauges"
Kleinheubacher Tagung, Miltenberg, Germany, September, 2016

15. **M. Elsobky**, G. Alavi, U. Zschieschang, M. Aghamohammadi, M. Hassan, H. Klauk, C. Harendt, H. Richter, J. N. Burghartz:
"Application of BCB-on-Si based substrate in combining flexible organic electronics and high frequency antenna"
International Symposium Organic Electronics, Stuttgart, Germany, June, 2016

Magazine Articles

1. B. Albrecht, **M. Elsobky**, T. Deuble, C. Harendt, J. N. Burghartz:
"The flexible package for ultra-thin sensors and readout chips"
Organic and Printed Electronics Journal (OPE), no. 29, pp. 14–15, November, 2019
2. **M. Elsobky**, Z. Yu, T. Deuble, J. N. Burghartz:
"MultiSense: Ultra-Thin Reconfigurable Voltage and Capacitance Readout ASIC"
EUROPRACTICE Activity Report, 2018–2019, pp. 23, January, 2019

Acknowledgments

Those who do not thank people do not thank God
—Muhammad PBUH

Many people have helped me in making this work possible. I thank them with sincere appreciation and gratitude.

First and foremost, I am very grateful to my advisor (Doktorvater) Prof. Dr.-Ing. Joachim N. Burghartz chairman of IMS CHIPS, for giving me the opportunity to work under his precious supervision. His invaluable vision and creative ideas made this work possible.

Furthermore, to Prof. Dr.-Ing. Jens Anders director of the Institute of Smart Sensors, University of Stuttgart, who provided me with his expertise, knowledge and encouragement, especially through the past two years, thank you. Without his constructive critiques and recommendations, this work would not have been the same.

Being at IMS CHIPS for the past four years, I have got the opportunity to meet many talented specialists. I would like to thank Dr. Christian Burwick and Ir. Cor Scherjon for their supportive advice and help during chip layout and digital design. I would like to extend my sincere appreciation to Dr. Christine Harendt and Mr. Thomas Deuble for their smooth management of KoSiF and ParsiFAl4.0 projects, which enabled in-part the funding and financial support of this work. A due word of thanks goes to Dr. Harald Richter and Dr. Zili Yu, for the continuous support and guidance, especially the valuable feedback and corrections of my thesis and publications.

At IMS CHIPS, I have also enjoyed the company of a tremendous office-mate Dr.-Ing. Moustafa Nawito, as he has always encouraged me to explore new possibilities. Furthermore, I am indebted to my friend Dr.-Ing. Muhammed Alshahed for the in-depth explanations of semiconductor physics. Succeeding him in teaching the tutorials of the ACDT course at the University of Stuttgart is not an easy lift. Additionally, I am glad to have gained several friends and would like to thank Dr.-Ing. Saleh Ferwana, Dr.-Ing. Yigit Mahsereci, Mr. Alexander Frank, Mrs. Fida Saidani,

Mrs. Ann-Christin Köllner and Mrs. Ulrike Paßlack for enabling a great work environment. To my bachelor, masters and HiWi students, Mrs. Bhavana Krishna Murthy and Mr. Amro Eldebiky in particular, thank you for your hard work and diligence in this early stage of your career.

I would like to acknowledge the vital role of the advanced packaging and device integration team at IMS CHIPS, particularly Dr. Albrecht Björn, Dr.-Ing. Mahadi Hassan and Dr.-Ing. Golzar Alavi, for their valuable support during the development of HySiF. I owe my profound gratitude to Mr. Stephan Schmiel and Mr. Dennis Kühnle for their support during chip assembly and wire-bonding.

I would also like to extend my acknowledgement to the project partners, notably Mr. Stefan Saller from Festo and Mr. Matthias Mahlich from Bosch. Moreover, I would like to acknowledge with much appreciation my friend Mr. Ayman Khattab of the Institute of Smart Sensors, University of Stuttgart, for introducing me to the modeling of $\Delta\Sigma$ converters.

For completing my B.Sc. degree, I have studied at the German University in Cairo (GUC) for five years, where I have met passionate and hardworking teachers. Among them are two wise scholars in Mathematics and Physics, namely Prof. Mustafa Amer and Assoc. Prof. Alaa Eldin Abdellatif, who keep on living in my memories. I will forever be grateful to the knowledge and wisdom that they shared.

Furthermore, a due word of thanks goes to Dr.-Ing. Yasmine Elogail for her continuous support throughout my academic degrees and notably our fruitful discussion in the field of semiconductor technology. I am also very fortunate to have completed my M.Sc. degree at the University of Ulm. It has been a life-changing period, as I have started my first steps in advanced IC design, which was guided by great scholars, such as Prof. Dr.-Ing. Maurits Ortmanns. To this end, I would like to express my gratitude to the German Academic Exchange Service (Deutsche Akademische Austauschdienst—DAAD) for granting me a scholarship during my M.Sc. studies. Moreover, I would like to acknowledge the financial support of the German funding agencies, BMBF for the projects KoSiF (#1612000461) and ParsiFAL4.0 (#16ES0435) and the State Ministry of Baden-Wüttemberg for Economic Affairs with the project BW-CPS (#AZ3-4332.62-HSG/77).

Last but not least, I must save my most intense gratitude and appreciation to my family and friends. Without their support, this journey would have been all the more difficult. To my beautiful and charming wife, Soheila, thanks for bringing the best out of me, your diligent character and sensible heart brighten my world. I would like to thank my beloved mother, who raised me and gave me a solid education and a strong character. I am sure that she keeps an eye on me, and I will always strive to make her proud. My sincere gratitude goes to my father, who always supports me in all my pursuits. To my caring and kind sister, Amira, thank you my dear for your endless support and motivating calls. To them all, I give my undying love and gratitude.

Contents

Nomenclature

Abbreviations

a-Si	Amorphous silicon
AAF	Antialiasing filter
AMOLED	Active matrix organic light-emitting diode
ASIC	Application-specific integrated circuit
BJT	Bipolar junction transistor
CDC	Capacitance-to-digital converter
CFP	Chip-Film Patch
CIFF	Cascaded integrator feedforward
CMFB	Common mode feedback
CoI	Chain of integrators
CTAT	Complimentary to absolute temperature
CTE	Coefficient of thermal expansion
CVC	Capacitance-to-voltage converter
DAC	Digital-to-analog converter
DBG	Dicing-before-grinding
DDA	Differential difference amplifier
DEM	Dynamic element matching
DNL	Differential nonlinearity
DR	Dynamic range
DSM	Deep submicron technologies
DT	Discrete time
e-skin	Electronic skin
ENOB	Effective number of bits
FD	Fully differential
FEM	Finite element method
FoMS	Schreier Figure-of-Merit
FoMW	Walden Figure-of-Merit
FPGA	Field programmable gate array

FSM	Finite-state machine
GBW	Gain-bandwidth
GO	Graphene oxide
HySiF	Hybrid Systems-in-Foil
I-ΔΣ	Incremental ΔΣ
IC	Integrated circuit
INL	Integral nonlinearity
IoT	Internet of Things
LCP	Liquid crystal polymer
LSB	Least significant bit
MCU	Microcontroller unit
MEMS	Micro-electromechanical systems
MSB	Most significant bit
NFC	Near-field communication
NS	Noise-shaping
NTF	Noise transfer function
OLED	Organic light-emitting diode
Op-Amp	Operational amplifier
OSR	Oversampling ratio
OTA	Operational transconductance amplifier
OTFT	Organic thin-film transistors
PCB	Printed circuit board
PDMS	Polydimethylsiloxane
PEN	Polyethylene naphthalate
PGA	Programmable gain amplifier
Poly-Si	Poly-crystalline silicon
PTAT	Proportional to absolute temperature
RFC	Recycling folded cascode
RFID	Radio frequency identification
RH	Relative humidity
RTD	Resistive temperature detector
S/H	Sample-and-Hold
SAR	Successive approximation register
SC	Switched capacitor
SNDR	Signal-to-noise+distortion ratio
SoC	System-on-Chip
SPI	Serial peripheral interface
SQNR	Signal-to-quantization-noise ratio
SR	Slew rate
STF	Signal transfer function
SWO	Single wire interface
TC	Temperature coefficient
TDC	Time-to-digital converter
TFT	Thin-film transistors
TG	Transmission gate

UTC	Ultra-thin chip
VCO	Voltage-controlled oscillator
VTC	Voltage transfer characteristics

Symbols

σ	Mechanical stress
ϵ	Mechanical strain
ρ	Resistivity
ν	Poisson ratio
ε_r	Dielectric constant
μ	Mobility
BW	Bandwidth
GF	Gauge factor
M	Oversampling ratio
Q	Quality factor
R_c	Contact resistance
R_{sh}	Sheet resistance
SR	Slew rate
V_{th}	Threshold voltage

Chapter 1
Motivation

For the last half-century, the rapid scaling and *cramming* of silicon devices in digital integrated circuits (ICs)Integrated Circuit have followed the well-known *Moore's law* [1]. To this end, reliable handling and processing of silicon wafers require thick and mechanically stiff substrates [2]. Nevertheless, silicon wafers and chips can be rendered ultra-thin and, thus, flexible, with a thickness of $20\mu m$ and below [2–4]. Fortunately, ultra-thin chips (UTCs)Ultra-Thin Chip are key enablers of another scaling degree of freedom in which the 3D stacking of chips allows for more compact system-in-package (SiP) solutions [2]. However, economical feasibility, structural reliability, elevated stress levels and altered behavior of the integrated devices are main challenges [5–7].

In contrast, large-area printed electronics, which is based on solution-processed amorphous semiconductor materials, offers an economic high-throughput alternative to UTCs. They can be deposited at room temperature and over large areas of a wide range of unconventional substrates, such as plastic, textiles and paper. For instance, AMOLED (active matrix organic light-emitting diode)Active Matrix Organic Light-Emitting Diode displays are increasingly demanded as a display for high-end wearables as they are lightweight, flexible and provide remarkable picture quality. However, life span, water resistance, high power consumption to display white color and current manufacturing cost are main challenges [8].

In summary, the advances in silicon technology have focused on the prime target of scaling and miniaturization, meanwhile, the progress in large-area electronics has promoted form flexibility and simpler, and, thus, faster manufacturing methods. Therefore, a hybrid integration, which encompasses silicon technology and large-area printed electronics into one single system, is a highly promising approach, known as Hybrid System-in-Foil (HySiF), for realizing high-performance flexible electronic systems [9].

M. Elsobky, *Ultra-Thin Sensors and Data Conversion Techniques for Hybrid System-in-Foil*, Recognizing Outstanding Ph.D. Research,
https://doi.org/10.1007/978-3-030-97726-9_1

1.1 Research Objectives

The main objective of this thesis is the implementation of HySiF-compatible electronic components, which, when integrated into polymeric foil, seamlessly operate within a high-performance flexible smart sensor system. More specifically, this work addresses the design, in part the fabrication, and the characterization of on-foil passives and thin-film circuits, in addition to microcontrollers, sensor readout and data converters implemented using silicon UTCs. However, this work does not cover the manufacturing or the integration technology of a complete HySiF.

For clarity, the terms "ultra-thin" and "thin-film" are used in this thesis to describe electronic systems with thicknesses less than 100 µm and 100 nµ, respectively. Also, the term "flexible" does not necessarily indicate an "ultra-thin" form factor and vice versa. However, most of the implemented ultra-thin components are fabricated on flexible substrates, although some are not characterized during bending. Besides, stretchable electronics, a branch of flexible electronics, is outside the scope of this thesis.

1.2 Thesis Organization

As shown in Fig. 1.1, this thesis is arranged into seven chapters, including this chapter and the concluding Chapter 7. Chapter 2 features an overview of ultra-thin flexible electronics. Besides, the HySiF concept is defined and its generic system-level model is presented. Furthermore, the UTC embedding technology, namely Chip-Film Patch (CFP), is briefly discussed.

Chapter 3 is devoted to the implementation of HySiF-compatible off-chip/on-foil passive and active components. Here, ultra-thin temperature, humidity and strain sensors are presented. Furthermore, flexible spiral inductors are introduced, which target several applications, including near-field communication, energy harvesting and passive LC sensing. Moreover, a large-area sensor addressing circuit is implemented using low-voltage organic thin-film transistors.

In Chapter 4, the focus is directed to implementing power-efficient sensor readout systems on silicon UTCs. Here, a HySiF-compatible ultra-thin humidity and strain-gauge sensor system is described. Furthermore, the effects of thinning on the performance of sensor readout and microcontroller ICs are investigated.

Chapter 5 provides an overview of the design space for flexible, as well as reconfigurable analog-to-digital converters (ADCs). A proposed technique for the power-efficient reconfiguration of incremental data converters is presented. The circuit design and measurement results of the proposed ADC before and after thinning are presented in Chapter 6.

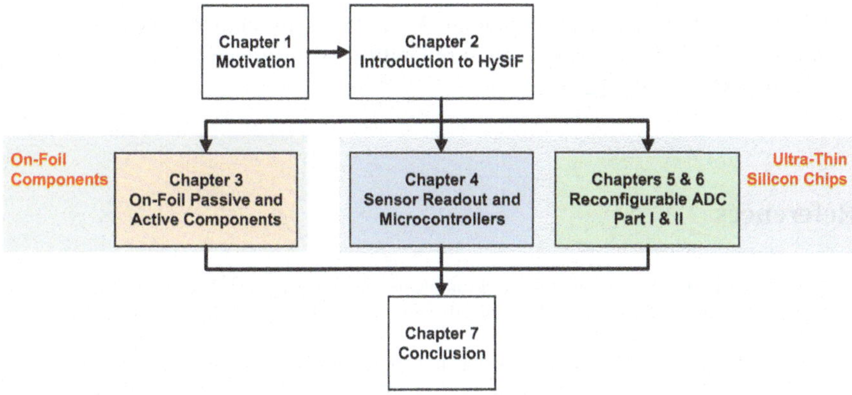

Fig. 1.1 Block diagram reflecting the organization of this thesis

1.3 Research Contributions

The main contributions of this work to advance the state-of-the-art, which have led to 1 book, more than 7 journal and 14 conference papers, are as follows:

- **Smart Ultra-Thin Humidity Sensor**—An ultra-thin humidity sensor, which is suitable for HySiF integration, is designed and fabricated considering the CMOS- and CFP-compatibility of its processing steps. Furthermore, a complete HySiF compatible sensor system is demonstrated by connecting 30-μm readout and microcontroller UTCs to the humidity sensor [10–14].
- **Ultra-Thin Temperature Sensor**—A comparative study of the electro-thermal behavior of thin-film platinum temperature sensors is presented. The resistive temperature sensors are fabricated on rigid and CFP-compatible foil substrates [13–15].
- **Flexible Spiral Inductors**—The design and electrical characterization of spiral inductors, which are fabricated on polymeric foil, enables the integration of near-field antennas, energy harvesting and passive LC sensing into HySiF [12, 14].
- **Smart Flexible Strain Gauge Sensor**—The combination of organic electronics, printed strain gauges and silicon UTCs has enabled the realization of electronic smart skin for robotic applications. The mentioned components are characterized after their integration into a single polymeric foil [14, 16, 17].
- **Reconfigurable ADC**—A novel implementation of incremental ADCs is proposed in this work. For the first time and to the best of the author's knowledge, that a resolution reconfigurable ADC is implemented on silicon UTCs and is able to span the 12–15 bits range in a power-efficient manner [18, 19].
- **Thinning Effects on CMOS Circuits**—The effects of the chip thinning on the implemented reconfigurable ADC, as well as on other readout and microcontroller ICs, have been investigated. This has been achieved in part by incorporating the

effect of stress in the circuit simulation. Additionally, the chips are measured before and after thinning to investigate their reliability and ensure proper operation before CFP embedding [10, 12–14].

References

1. Moore GE (2006) Cramming more components onto integrated circuits, reprinted from electronics, volume 38, number 8, April 19, 1965, pp. 114. IEEE Solid-State Circ Soc Newsl 11(3):33–35
2. Burghartz JN (2011) Ultra-thin chip technology and applications. Springer, New York
3. Burghartz JN (2013) You can't be too thin or too flexible, vol 50. Spectrum, IEEE, pp 38–61
4. Zimmermann M, Burghartz JN, Appel W, Remmers N, Burwick C, Wurz R, Tobail O, Schubert M, Palfinger G, Werner J (2006) A seamless ultra-thin chip fabrication and assembly process. In: 2006 International electron devices meeting, Dec 2006, pp 1–3
5. Hassan M, Angelopoulos EA, Rempp H, Endler S, Burghartz JN (2010) Packaging challenges associated with warpage of ultra-thin chips. In: 3rd Electronics system integration technology conference ESTC, Sep 2010, pp 1–5
6. De Munck K, Chiarella T, De Moor P, Swinnen B, Van Hoof C (2008) Influence of extreme thinning on 130-nm standard CMOS devices for 3-D integration. IEEE Electron Device Lett 29(4):322–324
7. Takyu S, Kurosawa T, Shimizu N, Harada S (2006) Novel wafer dicing and chip thinning technologies realizing high chip strength, vol 970, p 5
8. Koden M (2017) OLED displays and lighting: flexible OLEDs. IEEE. Available https://ieeexplore.ieee.org/document/7827447
9. Burghartz JN, Appel W, Harendt C, Rempp H, Richter H, Zimmermann M (2009) Ultra-thin chips and related applications, a new paradigm in silicon technology. In: 2009 Proceedings of the European solid state device research conference, Sep 2009, pp 29–36
10. Elsobky M, Albrecht B, Richter H, Burghartz JN, Ganter P, Szendrei K, Lotsch BV (2017) Ultra-thin relative humidity sensors for hybrid system-in-foil applications. In: 2017 IEEE Sensors, Oct 2017, pp 1–3
11. Elsobky M, Albrecht B, Ganter P, Szendrei K, Deuble T, Harendt C, Lotsch B, Burghartz JN (2017) Relative humidity sensors for system-in-foil applications. In: MikroSystemTechnik 2017; Congress, Oct 2017, pp 1–4
12. Elsobky M, Alavi G, Albrecht B, Deuble T, Harendt C, Richter H, Yu Z, Burghartz J (2018) Ultra-thin sensor systems integrating silicon chips with on-foil passive and active components. In: Proceedings, vol 2, p 748
13. Elsobky M, Deuble T, Ferwana S, Albrecht B, Harendt C, Ottaviani A, Alomari M, Burghartz JN (2020) Characterization of on-foil sensors and ultra-thin chips for HySiF integration. IEEE Sens J 1–1
14. (2020) Chip-flex technology: hybrid systems-in-foil. In: Elements in flexible and large-area electronics. Cambridge University Press
15. Elsobky M, Ottaviani A, Alomari M, Yu Z, Deuble T, Burghartz JN (2019) Characterization of thin-film temperature sensors and ultra-thin chips for HySiF integration. In: 2019 IEEE international conference on flexible and printable sensors and systems (FLEPS), July 2019, pp 1–3
16. Elsobky M, Mahsereci Y, Yu Z, Richter H, Burghartz JN, Keck J, Klauk H, Zschieschang U (2018) Ultra-thin smart electronic skin based on hybrid system-in-foil concept combining three flexible electronics technologies. Electron Lett 54(6):338–340
17. Elsobky M, Mahsereci Y, Keck J, Richter H, Burghartz J (2017) Design of a CMOS readout circuit on ultra-thin flexible silicon chip for printed strain gauges. In: Advances in radio science, vol 15, pp 123–130

18. Elsobky M, Mohamed A, Deuble T, Anders J, Burghartz J (2020) A 12-to-15 b 100-to-25 kS/s resolution reconfigurable, power scalable incremental ADC. Express Briefs, Submitted, IEEE Transactions on Circuits and Systems II
19. Elsobky M, Anders J, Burghartz J (2020) A resolution reconfigurable incremental ADC with scaled ring-amplifier-based integrator. Electron Lett (Submitted)

Chapter 2
Introduction to Hybrid System-in-Foil

Hybrid System-in-Foil (HySiF) is an emerging approach that targets the economical integration of multiple electronic components into polymeric foil. This chapter starts with an overview of flexible electronics while highlighting the recent advances in integration technologies. Furthermore, the HySiF concept is defined and its passive and active components are discussed. To this end, a generic model for a smart sensor system reflecting the HySiF concept is presented. Finally, the fabrication and electrical characterization of the Chip-Film Patch (CFP) technology, which is an integral part of the HySiF outlined in this work, are presented.

2.1 Ultra-Thin Flexible Electronics

Plastic electronics, including flexible, ultra-thin, printed and organic, is an emerging field in which conformable electronic systems are designed to introduce intelligence and connectivity to almost every *Thing*. Consequently, flexible electronics is considered one of the main enablers of the Internet of Things (IoT). When combined with new materials and advanced fabrication processes, flexible electronics offer unique characteristics such as mechanical flexibility, thin-form factor, large-area scaling feasibility and adaptability to curved surfaces. In contrast to the dense and highly integrated devices on silicon chips, large-area electronics include devices that can be widely spread across a large substrate [1].

Thanks to wearable electronics, foldable phones and rollable displays, the market for consumer flexible electronics is rapidly expanding. For instance, the company NXP Semiconductors develops contactless chip modules targeting flexible electronic applications, such as e-health IDs, e-vehicle licenses and e-passports [2]. The company's progress toward thickness reduction is notable as it currently reached a total IC package thickness of 200 μm. Furthermore, the company Infineon Technologies AG implements smart card payment solutions and is currently developing a semi-flexible

M. Elsobky, *Ultra-Thin Sensors and Data Conversion Techniques for Hybrid System-in-Foil*, Recognizing Outstanding Ph.D. Research,
https://doi.org/10.1007/978-3-030-97726-9_2

system-on-card demonstrator aiming at combining higher computational power and security with advanced fingerprint identification [3].

In the field of health care, personalized and predictive medical care are enabled by the emerging soft biosensor systems that can provide critical insights into health status by monitoring vital biomarkers in real-time [4]. Furthermore, electronic skin (E-skin) patches, such as smart bandages, cardiac monitoring devices and drug injectors, have already established over \$7.5 bn in revenue in 2018 with a growth forecast of \$20bn per year by 2029 [5].

2.1.1 Integration Technologies

The well-established Printed Pircuit Board (PCB) manufacturing has inspired the realization of various integration and packaging technologies for ultra-thin and flexible electronic components. The natural extension of the rigid PCBs are the flexible PCBs, in which the brittle FR4 substrate is replaced with flexible sheets of polymers, such as Polyimide-based resin [6] and Liquid Crystal Polymer (LCP)[7]. Board thicknesses from 1mm down to 100 μm are achieved depending on the number of metal layers and substrate material. Since flexible PCBs integrate conventional through-hole and surface mount components in a similar fashion to the rigid PCBs, their flexibility and form factor are limited by the utilized active and passive components.

Another variant of flexible PCBs uses bare die UTCs, which are either embedded in the substrate sheets or mounted on the substrate surface using flip-chip bonding [8]. The form factor is significantly reduced but another challenge remains, which is to achieve fine-pitch circuit interconnects, particularly the fan-out interconnects of UTCs with large I/O count. Current flexible PCB manufacturing technologies provide a minimum pitch in the range of 100 μm, which is limited by the metal structuring and via drilling processing steps. This, in turn, restricts the pin-count for the utilized ICs and reflects the need for an alternative integration technology.

In order to fill this gap, ultra-thin chip embedding technologies, such as Ultra-Thin Chip Package (UTCP) [9], Chip-Film Patch (CFP) [10, 11] and Flex Silicon-on-Polymer [12], have emerged to provide a conformable UTC package with fine-pitch wire interconnects. In these technologies, spin-on polymers (e.g. PI and BCB), which are commercially available in liquid form, are used to provide more control on the package composition and the thickness of each layer (package thickness less than 100 μm). Conventional CMOS-compatible lithography is used to achieve structures down to 10 μm at the expense of lower manufacturing throughput. Furthermore, other passive and active flexible electronic components can be manufactured alongside the embedded UTC(s) leading to a system-in-foil [1, 13–15]. In other applications, the UTC fan-out package can be integrated as an interposer into another flexible PCB, which removes the typical limitations of flexible PCBs and raises its added value [16, 17].

For manufacturing thin-film and organic electronics (thickness less than 100 nm), liquid printing techniques, such as screen, gravure, offset and inkjet printing, are

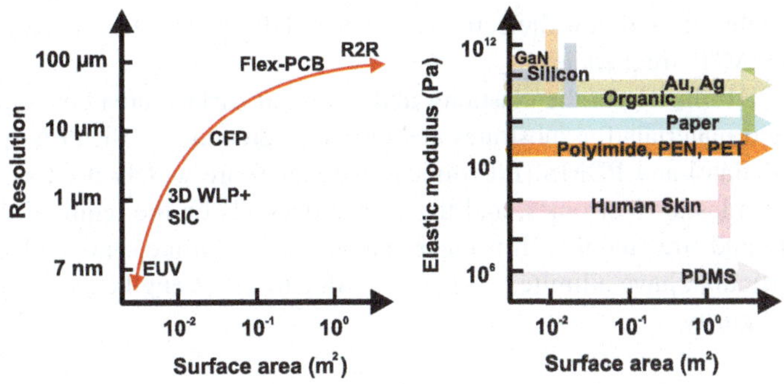

Fig. 2.1 Graphical illustration of the surface area coverage versus. **a** the resolution capability of various lithography and integration technologies, as well as **b** the elasticity of different substrate and circuit interconnect materials. Selected lithography techniques, such as extreme ultra-violet (EUV) lithography and R2R processing are chosen to illustrate the general trend. Ultra-thin chips and wafers are essential elements for the advances in 3D stacked ICs (SIC) and 3D wafer-level packaging (WLP)

widely used [18]. Additionally, roll-to-roll (R2R) manufacturing have enabled the mass production of large-area electronics that incorporate basic circuits (e.g. thin-film solar cells [19, 20] and Organic Light-Emitting Diode (OLED) for lighting and displays [21, 22]). Figure 2.1a compares the resolution capability vs. the area coverage of selected lithography techniques, such as extreme ultra-violet (EUV) lithography and R2R processing.

The full utilization of R2R processing can be achieved when ultra-thin and flexible electronic components incorporating high-performance complex circuitry (e.g. microprocessors and data converters) are coherently integrated into the R2R processing, and, thus, smart and high-value large-area electronic products can be attained. In fact, CFP is considered a kind of sheet-to-sheet (S2S) process technology. In this context, mix-and-match of UTC embedding (e.g. CFP) and R2R processing is one possibility for achieving the HySiF concept. However, careful study of material properties and boundary conditions for temperature and residual stress are necessary to achieve a proper heterogeneous integration [1, 23].

2.1.2 Substrate Materials

The electrical, thermal and mechanical properties of the substrate and packaging material significantly impact the performance of the assembled ultra-thin flexible micro-systems. For instance, the mismatch in the coefficient of thermal expansion (CTE) between the substrate and assembled micro-system induces interface residual stress during the fabrication process (e.g. R2R [23] and CFP [24, 25]). Besides, substrate materials with low dielectric constant and dissipation factor are needed for beyond 100 MHz frequencies.

Figure 2.1b illustrates the elasticity and potential surface area coverage of frequently used materials for substrates and circuit interconnects. PI, PEN (polyethylene naphthalate) and PDMS (Polydimethylsiloxane) are widely used as substrate materials in plastic electronics, and the choice depends on the required thickness, bendability and stretchability. It is clear that single-crystalline semiconductors, silicon in particular, have a limited surface area due to the strengthened requirements on material purity.

2.2 Hybrid System-in-Foil

The term hybrid usually describes the outcome of combining multiple elements, which were prefabricated using different technologies. The result of this combination is always thought to be more beneficial than the original single elements. This is not always true since the outcome of hybridization could serve as a transitional stage to new developing technology. For instance, at the beginning of the 21st century, the term hybrid becomes related to the effort of the car industry to switch from fossil fuel to the presumed cleaner electrical energy by manufacturing hybrid cars that operate using both fuel systems. Besides, the term hybrid is heavily used in the microelectronics field when two or multiple technologies are combined, for instance, the hybrid III–V/silicon technology and hybrid data conversion techniques.

Hybrid System-in-Foil (HySiF) is defined as an approach for a promising technology direction, which describes the fabrication process and design rules for integrating electronic components into polymeric foil. HySiF serves as a platform, which complements the merits of the Very Large Scale Integration (VLSI) of silicon ICs with the large-area electronics by combining mechanically flexible System-in-Packages and System-on-Packages (SiPs and SoPs) concepts.

Figure 2.2 graphically illustrates one variant of HySiF as it includes, large-area and on-chip sensors, silicon-based and non-silicon-based ICs, antennas, passives, display and energy storage elements [1, 14, 17, 26, 27]. Here, UTCs are main building blocks, which are either directly embedded in the polymeric substrate or indirectly assembled via an interposer package. In this section, design and fabrication considerations for commonly used active and passive components (cf. Fig 2.3) are presented to enable their integration into and compatibility with HySiF.

2.2.1 Active and Passive Components

Since HySiF targets an ultra-thin form factor regardless of the target surface area, structural and geometrical restrictions are applied to the electronic components that are incorporated into HySiF. For instance, conventional surface mount passives are

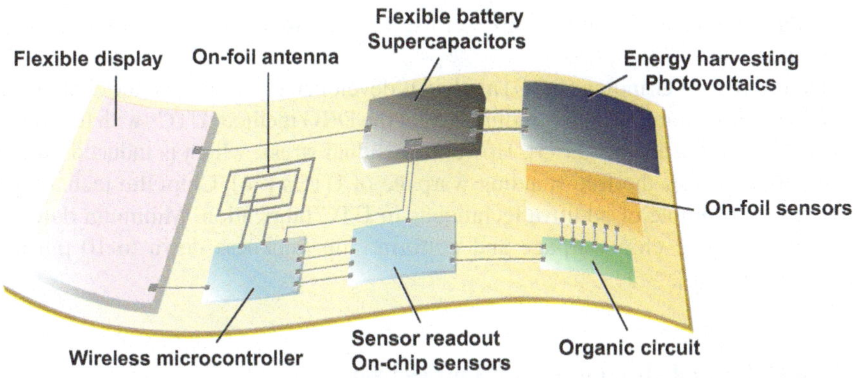

Fig. 2.2 Concept of a Hybrid System-in-Foil (HySiF) [1, 14, 17, 26] in which multiple ultra-thin silicon ICs are combined with large-area electronic components and integrated into flexible polymeric foils

not compatible with the presented HySiF. Therefore, all the HySiF components are designed to be self-contained, i.e. during their operation, they do not depend on external auxiliary components (e.g. decoupling capacitors, matching resistors and clamp diodes). For instance, certain wired communication protocols, such as I^2C, require pull-up or pull-down resistors for their proper operation. From the HySiF perspective, such resistors should be implemented on-chip to avoid the need for extra off-chip/on-foil components. If needed, thin-film passives can be fabricated directly on the foil surface using standard lithography [28, 29], screen printing [30] or inkjet printing [31]. However, they introduce additional process steps to the HySiF fabrication and their quality does not match the conventional ones.

Ultra-Thin Chips

Since the active part of the chips takes about 1% of its total thickness, chip thickness is a degree of freedom independent of conventional scaling. Traditionally, UTC fabrication is divided into subtractive and additive techniques [32, 33]. Without loss of generality, subtractive techniques are simpler to fabricate UTCs, though they cause material waste, which is difficult to be recovered [34]. On the contrary, additive techniques optimize material usage to achieve economic UTC fabrication.

Silicon-on-insulator (SOI) wafers can be utilized as a substractive method in the UTC fabrication. The buried oxide acts as an etch stop layer which results in a precise definition of the chip thickness. However, boundaries on the technology choice, development costs and thin wafer handling are main concerns. Wafer back-thinning/back-grinding is another technique to fabricate UTCs, which is widely used by semiconductor manufactures to address the increasing demand for a compact and ultra-thin form factor. However, the mere adoption of back-grinding is limited by the degradation in the structural integrity of the thin wafer/chip as micro cracks start to propagate.

Dicing-before-Grinding (DBG) method is developed to circumvent the challenges associated with conventional grinding. Using the DBG method, UTCs with thickness down to 20 μm is achieved [35]. However, residual stress, which is induced during grinding, increases the free-standing warpage of UTCs [36]. ChipFilm technology is a known example of additive techniques in UTC fabrication. Minimum defects, low free-standing chip warpage and uniform chip thickness down to 10 μm are achieved [32, 34].

Metal-Oxide and Organic Thin-Film Electronics

A new paradigm in flexible electronics is enabled by the field of thin-film electronics, as it offers inherently thin and bendable devices, as well as a simple, room-temperature and quick fabrication process with potentially less environmental impact when compared to the silicon technology. Within the HySiF framework, thin-film electronics, including metal-oxide and organic electronics, complement and operate alongside the mature silicon UTCs. Due to the amorphous nature of metal-oxide and organic semiconductors, they can be fabricated on a wide range of substrates [37]. However, their thin-film and semiconducting properties are inevitably impacted by the surface roughness of the flexible substrate. For instance, organic Thin-Film Transistors (TFTs) fabricated on PEN substrate has achieved higher carrier mobility (2.7 $cm^2/(Vs)$) compared to those fabricated on paper (0.8 $cm^2/(Vs)$) [38, 39].

Using TFTs, sensor array addressing and signal multiplexing is no longer limited by the I/O count of the silicon ICs. For instance, a large-area array of frequency-hopping ZnO TFT oscillators have been implemented in [40], in order to extend the number of available sensor channels from the typical quadratic behavior (N^2) of active matrix addressing to an exponential behavior (2^N).

In the field of neuroscience, particularly for in vivo intracellular and extracellular recordings, OLEDs and TFTs have been integrated into the tip of multi-electrode arrays for photostimulation and signal amplification, respectively [41–43]. Here, TFTs offer high interface capacitance, which results in high signal-to-noise ratios [43].

Circuit Interconnects

For circuit interconnects, gold (Au), silver (Ag) and aluminum (Al and AlSiCu) are commonly used metals. The thinner the metal, the more flexible it becomes but at the same time the sheet resistance increases. This results in an additional RC delay, which remains a challenge for signal integrity in high-speed flexible micro-systems. As circuit interconnects get longer, the value of the parasitic capacitance increases. Beside limiting the signal speed, such parasitic capacitance increases the power consumption and places strengthened requirements on the sensor instrumentation [47].

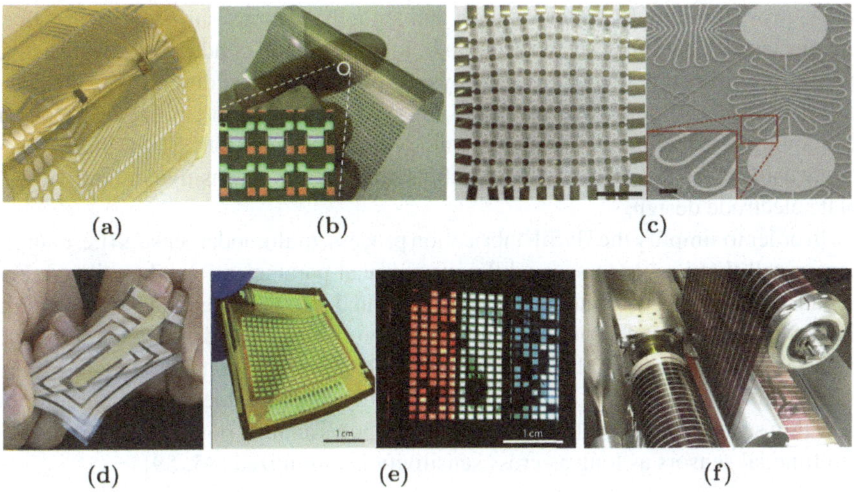

Fig. 2.3 Selected examples of ultra-thin flexible passive and active components. **a** Two 30-μm UTCs embedded in CFP [44]. **b** Array of organic TFTs fabricated on 20-μm thick paper substrate (area = $70 \times 70\,\mathrm{mm}^2$). Reproduced from [37] with permission from John Wiley and Sons. **c** Skin-inspired network of vertically stacked temperature, humidity, in-plane strain, proximity, UV light and magnetic field sensor arrays fabricated on PI substrate [45]. **d** Screen-printed antenna and sensor, which are integrated into a stretchable sticker that adheres to skin and tracks human pulse [46]. **e** Flexible full-color AMOLED display using carbon nanotube TFT backplane [22]. **f** R2R manufacturing of organic photovoltaics [20]

Moreover, the electrical behavior of thin-film metal interconnects deviates from bulk metal due to an improved piezoresistive effect [48, 49], which couples the RC delay of wire interconnects to the mechanical activity of the substrate.

In the presented HySiF variant, coplanar circuit interconnects are used. Since metals have higher Young's moduli compared to polymers, the density and layout of wire interconnect impacts the free-standing warpage of the HySiF. In order to improve the elasticity of interconnects, various layout techniques have been reported, such as zigzag and serpentine designs [50–52], but this is usually done at the expense of longer wires. To this end, an intrinsically elastic interconnects can be achieved using amorphous metals [53], Carbon Nanotube (CNT) [54] or even transparent Silver Nanowires (Ag NW) [55, 56]. However, the conductivity of conventional metals is still superior. Besides, the neutral plane of stress within the HySiF foil can be leveraged to improve the micro-mechanical reliability of critical components and circuit interconnects (e.g. CMOS chip fan-out) [1, 57].

Ultra-Thin Sensors

Sensors are integrated into HySiF while considering cumbersome boundary conditions regarding sensor protection, thermal and moisture effects of the next processing steps, providing low RC interconnects and smooth substrate surface. For instance, the electrical properties (i.e. electron/hole, ionic or proton conductivity) of resistive thin-film sensors are affected by the physical properties, in addition to the mechanical activities, of the flexible substrate. This correlation can be mitigated by either employing bridge configurations, special layout techniques [50, 51] or more effectively using capacitive sensors that benefit from the stress-insensitivity of the parallel plate electrode design.

In order to simplify the HySiF fabrication process, multimodal sensors (i.e. a single sensor cell that is able to sense different physical parameters) can be utilized. For instance, a ferroelectric gate insulator is integrated into the gate oxide of an organic TFT [58]. Since ferroelectricity includes both pyroelectricity and piezoelectricity, and by establishing a tensor relation for the TFT electrical parameters, the temperature and mechanical strain can be simultaneously detected. Besides, 3D integration of thin-film sensors by vertical stacking (cf. Fig. 2.3c) is another possibility of achieving multimodal sensors as long as cross sensitivity is minimized [45, 59].

Flexible Antennas

The electrical behavior of antennas, flexible antennas in particular, is strongly coupled to the electromagnetic properties of the substrate, as well as to the passivation superstrate. Low electrical permittivity and dielectric loss in the vicinity of the antenna tend to improve its efficiency. PI and PDMS are widely used substrate materials with dielectric constants and loss tangent ranges of 3.4 ± 0.2, 0.0022 ± 0.0004 and 2.5 ± 0.2, 0.0017 ± 0.0003, respectively. Several flexible antennas have been reported and their performance is approaching their rigid counterparts [60–62]. For ultra-thin wireless systems, connection loss or signal distortion is expected when those systems are attached to metallic objects and batteries. To resolve this issue, a ferrite sheet or a combination of flexible metamaterials and ferrite thin-films can be integrated into the substrate [63, 64].

Flexible Displays

For ultra-thin flexible displays, the market is currently expanding as foldable portable electronics, notably smartphones, are on the rise. The Chinese company Royole is currently mass producing AMOLED flexible displays, which are light weight, 100-μm thick and can be bent downto 1 mm radius. However, the economic feasibility of integrating flexible displays into HySiF systems, such as smart labels and e-skins, is still in question.

Energy Harvesting and Storage

For autonomous and self-powered HySiF, local energy storage is needed. On the bright side, energy harvesting devices, such as photovoltaics [19, 20, 65] and electromagnetic field harvesters, can be rendered ultra-thin and flexible. On the other side, the form factor of energy storage devices is considered a bottleneck for the development of a portable seamless HySiF. Flexible lithium-ion batteries have been reported [66, 67]. However, sizeable form factor, low power density and heat generation are the main concerns. Complementary storage devices, such as flexible supercapacitors [68], can be used to supply high power demands but for limited duration.

2.2.2 *System-Level Concept*

Figure 2.4 shows a block diagram for a generic model reflecting the envisioned HySiF concept in which main electronic subsystems are defined. By closely considering a target application, this block diagram can be customized and certain blocks can be omitted (removal of flexible display for economic implementation). The colored blocks in Fig. 2.4, such as off-chip/on-foil sensors, organic circuits and data converters, are addressed within the scope of this work.

In the presented HySiF concept, a centralized System-on-Chip (SoC) UTC performs accurate sensor readout and data conversion. In addition, high-speed digital signal processing, wired and wireless communication and power management are incorporated. In this way, the complex electronic tasks are assigned to the well-established CMOS technology and the number of embedded silicon UTCs is limited to 2 or 3 ICs per HySiF.

Moreover, large-area off-chip sensor arrays are distributed on the foil as they complement the operation of the miniaturized on-chip smart sensors. Here, on-foil sensors require signal conditioning circuits, which eventually brings the signal flow back to the centralized SoC. Optionally, thin-film transistors can be employed to preamplify and multiplex the analog signals. Therefore, signal integrity is preserved and the number of on-foil sensors can be extended beyond the limits of the number of SoC pads. Wireless microcontroller UTCs are a good candidate for combining high-speed computational power with multi-standard wireless connectivity. For ultra-low power applications, bespoke microcontrollers on plastic substrates are good alternatives since their implementation is optimized for the target application [69].

2.2.3 *On-Chip Versus Off-Chip/On-Foil Components*

HySiF components can either be integrated into thin silicon chips or be fabricated directly on the foil substrate [1]. For sensing elements, Table 2.1 summarizes key

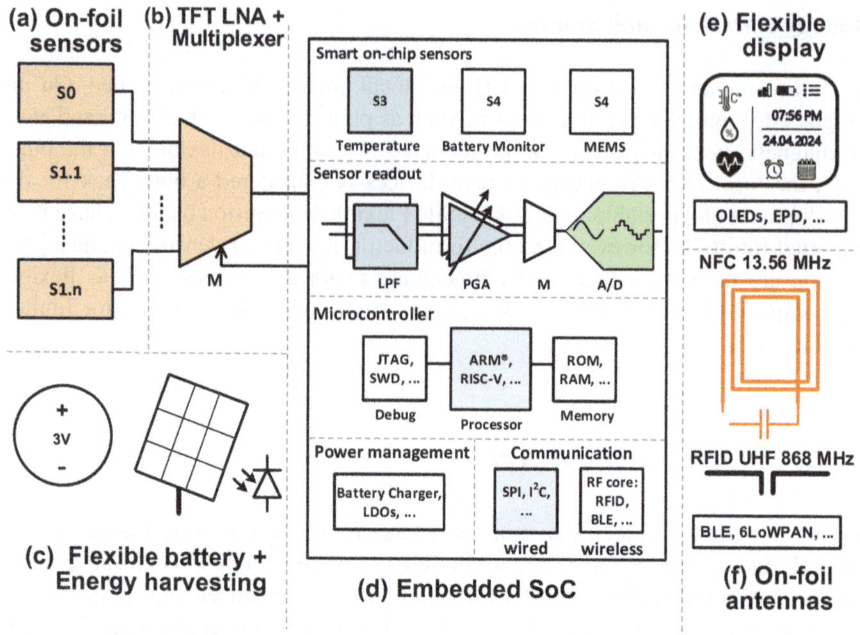

Fig. 2.4 Block diagram for a flexible generic model of a smart sensor system reflecting the targeted HySiF concept. The colored blocks, such as off-chip/on-foil sensors, organic circuits and data converters, are addressed within the scope of this work

differences between the concepts of integrated on-chip vs. large-area off-chip/on-foil sensors. It is clear that matrix or array formation is simpler and more economic for the on-foil sensors in contrast to the complexity of handling and assembling several silicon chips. Furthermore, signal integrity and readout speed are negatively impacted when long wire interconnects are used to connect the large-area sensors with a central readout IC.

For economic reasons, UTC embedding in CFP is limited to a maximum number of 2 to 3 chips per foil [1, 17]. The interface required between the large-area components and silicon UTCs limits the number of accessible on-foil sensors [40]. Figure 2.5a shows a schematic illustration of a core-limited pad ring UTC, which is embedded in CFP foil having larger perimeter fan-out on-foil pads. Here, the chip doesn't require high I/O count and the UTC-foil interconnection profits from such arrangement by spreading the UTC pads to the perimeter foil pads using a low resistance fan-out [24]. On the contrary, due to CMOS-scaling and increased SoC integration, the pad-limited pad rings are increasingly used. Consequently, the number of I/O pads increases, which demand more on-foil fan-out pads, as illustrated in Figure 2.5b. In this situation, the foil area in close proximity to the embedded UTC is utilized and minimum width of the wire interconnects are used for routing.

Table 2.1 Comparison between the concepts of on-chip versus off-chip/on-foil sensors

	On-chip sensors	On-foil sensors
Array formation	complex	simple
Interconnect parasitics	low	high
Readout electronics	integrated	remote
Signal integrity	☺	😐
Area coverage, footprint	small	large
Thickness	10 μm - few mm	<1 μm
Thermal contact resistance	high when in package	low
Lithography	complex	simple
Fabrication temperature	high	low
Power consumption	low	high
Reliable lifetime	years	weeks - months
Biodegradable	☹	☺
Substrate	mainly silicon	PI, BCB, LCP, paper, ...
Package	ceramic, plastic, ...	PI, BCB, LCP, ...

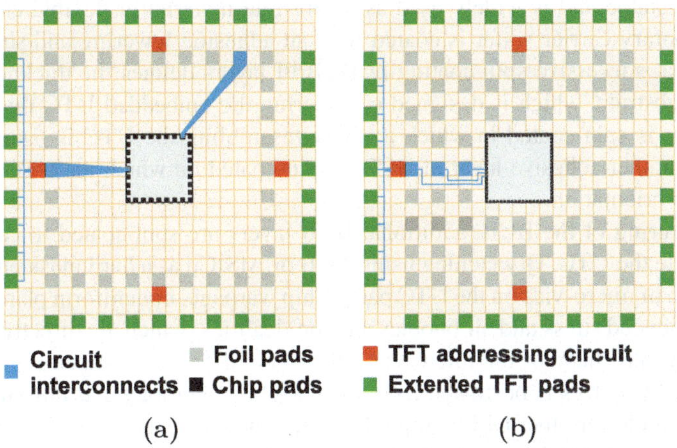

Fig. 2.5 I/O count extension using on-foil TFT sensor addressing circuits. **a** Core-limited and **b** pad-limited I/O pad ring design for embedded chips with the associated fan-out perimeter and area pads, respectively

For both pad ring variants and in order to address more on-foil sensors using the limited UTC access points, the chip I/O count can be extended using active TFT multiplexing circuitry. Different system architectures for large-area acquisition and I/O count extension are available, such as active matrices, binary addressing circuits and frequency-hopping oscillators [40]. In this thesis, a flexible low-voltage OTFT-based addressing circuit is developed, which incorporates a shift register and analog switches [15].

2.3 Ultra-Thin Chip Packaging: Chip-Film Patch

The CFP is an embedding technology, which provides a flexible fan-out package for single or multiple UTCs. The fine-pitch interconnects ($<10\,\mu m$) of the CFP enable the integration of high-performance pad-limited ICs into flexible electronic systems. Moreover, several forms of the HySiF concept are achieved in part using CFP technology. In fact, the CFP-compatibly is a principle aim for the ultra-thin active and passive components that are implemented in this thesis. Therefore, it is reasonable to overview the process sequence of one version of the CFP technology.

2.3.1 Fabrication Process Steps

Figure 2.6 shows the cross-section of the CFP processing steps [10, 24]. The flexible substrate is a stack of two spin-on polymers, namely BCB and PI [10, 24]. The embedding polymer is BCB, which is CMOS-compatible and has fine-pitch patterning capability. On the other side, PI is the reinforcing polymer, which is highly bendable, stable and is widely used as a substrate material in flexible electronics. The temporary carrier wafer is coated with an adhesion lowering and stress-relief layer. Next, successive spin-coating of BCB/PI layers defines the thickness of the flexible substrate, which is designed considering the embedded UTC thickness. A BCB layer is spin-coated in which a cavity resembling the UTC is patterned and etched. Another adhesive layer of BCB is spin-coated on which the UTC is placed in face-up orientation.

After curing of the BCB, additional BCB layers are spin-coated to completely encapsulate the UTC. Dry etching using BCB or AlSiCu as a hard mask enables the realization of micro-vias on the UTC pads. Chip warpage, nonuniform planarization, process-induced stress and air bubbles are nonidealities, which are directly linked to the quality and shape of the micro-vias. To counteract these nonidealities, the idea of standard via sizes is borrowed from the CMOS technology and introduced here as a design rule for other CFP-compatible components.

Conventional AlSiCu sputtering is used to fabricate the fan-out interconnects on the smooth BCB surface. A superstrate layer, with a thickness similar to the substrate thickness, is fabricated using successive spin-coating of BCB/PI layers. Finally, coarse fan-out pads are opened and optionally covered with a noble metal. When the CFP is used as an interposer, the CFP is mechanically released from the rigid carrier either using a manual or laser cutter. Alternatively, further processing steps are followed (cf. temperature and humidity sensors fabrication, discussed in chapter 3), when the CFP is utilized as a substrate for HySiF integration.

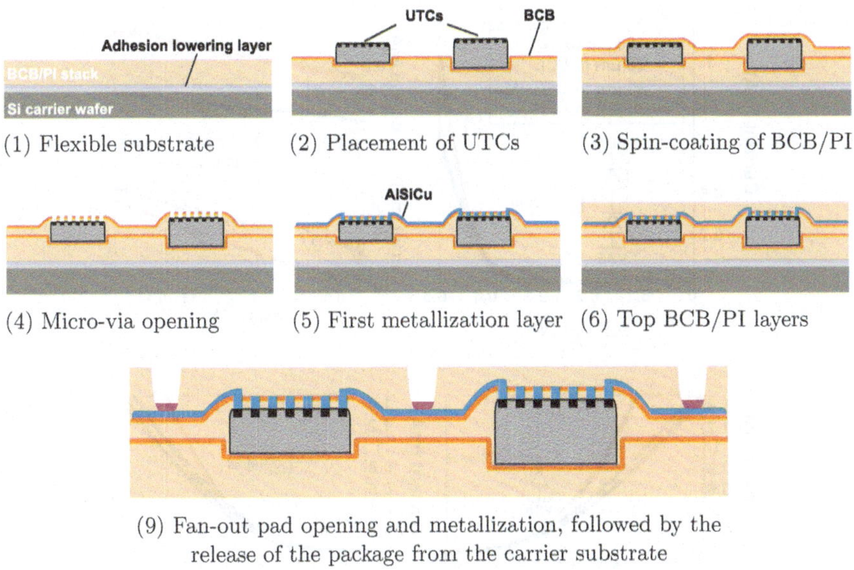

Fig. 2.6 Fabrication process steps for the CFP embedding technology [10, 24]

2.3.2 *Electrical Characterization: Continuity Test*

NMOS and PMOS devices, connected in diode configuration, are usually used in electrostatic discharge (ESD) protection circuits, which are placed near I/O, supply and ground chip pads. Here, such ESD devices are characterized in the presented continuity test, which checks the basic functionality of the embedded UTC(s) after critical CFP processing steps [17]. The measurement setup consists of a manual probe station and source measurement units (SMUs) (cf. Fig. 2.8a).

Figure 2.7 shows the measurement results of the ESD devices integrated into the Apollo microcontroller, which includes 4 sets, each containing 4× chip samples. The first and second sets correspond to 400- and 30-μm thick chips assembled in a conventional ceramic package. The third set is for the 30-μm UTCs during the CFP processing after the first metallization. The fourth is for the 30-μm UTCs when it is completely embedded after the final CFP process step.

The digital ground pad VSS is connected to ground while the digital supply voltage pad VDD is swept from 0 to −0.5 V. Consequently, multiple ESD devices are activated and the total current is plotted on Fig. 2.7a. This preliminary test checks the overall state of the chip supply and ground pins connections to the fan-out pads but does not say much about the performance of the embedded UTC or the connection of other I/O pads. Higher leakage current and lower on-current are detected in the measurements of the ESD devices in the embedded UTCs, which trace back to high stress and temperature present during the processing of the CFP.

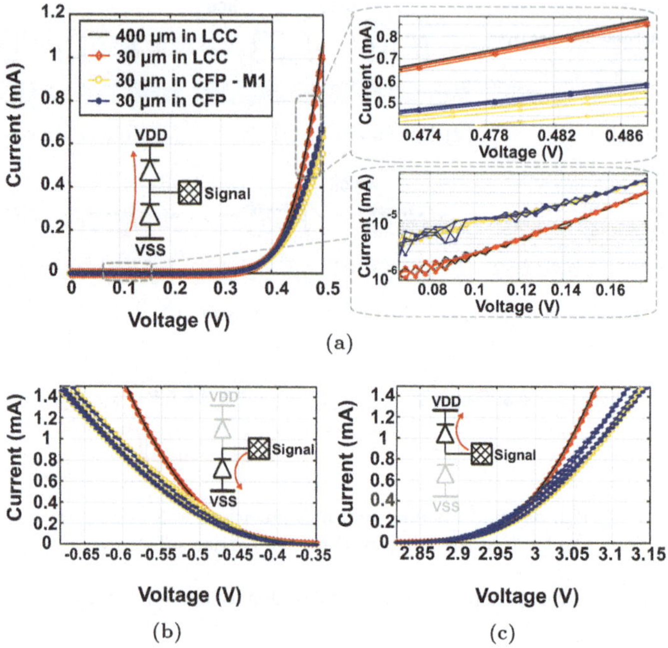

Fig. 2.7 Diode characteristics for the ESD devices of an embedded UTC in the CFP of Fig. 2.8b [17]. **a** Diode characteristics between digital supply and ground pads while highlighting the leakage and on-currents. Diode characteristics for a certain I/O pad while activating, **b** NMOS and **c** PMOS ESD devices, respectively

In Fig. 2.7b, c, the VSS and VDD pads are connected to ground and 2.5 V, respectively. The voltage level of a general I/O pad is swept from −0.8 to 3.3 V. Consequently, only the associated ESD devices (PMOS and NMOS diode-connected devices) of this I/O pad are successively activated. By repeating the continuity test on other I/O pads, the UTC embedding in the CFP is electrically validated.

It is interesting to observe that for NMOS ESD devices, the current curves do not vary after the CFP's first metallization and after the CFP final processing step. However, for PMOS ESD devices, the on-current increases after the CFP final processing step. This could indicate a slight stress relief on the embedded UTCs after its total encapsulation in the CFP, which occurs in the preferred direction of PMOS devices.

2.3.3 Smart Label Demonstrator

The first generation of smart labels was enabled by the RFID (Radio Frequency Identification) technology, in which a compact silicon chip with 2–4 pads is connected to a printed antenna. The reader device takes over the interrogation task and

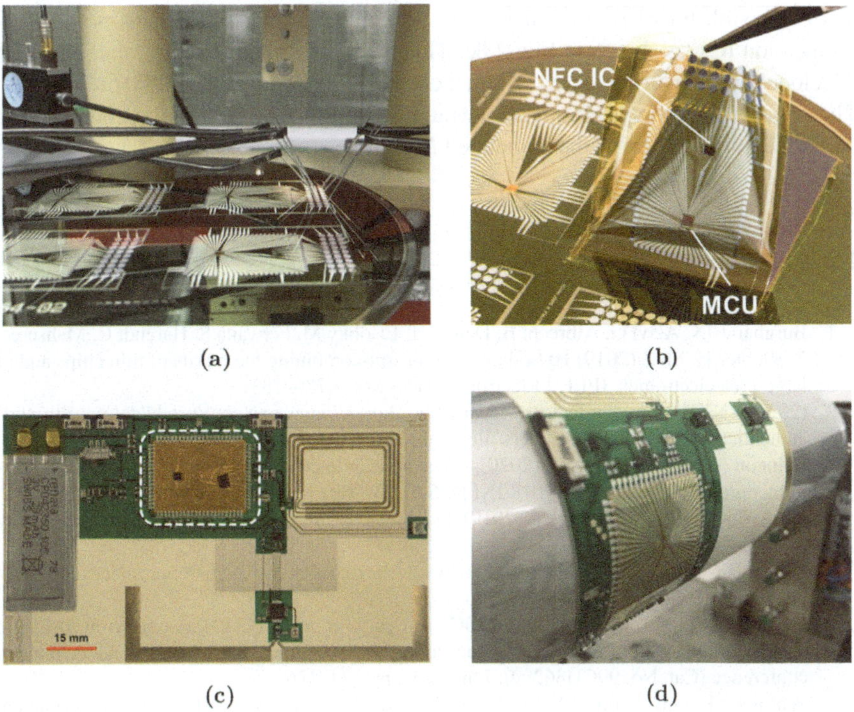

Fig. 2.8 High-performance smart label, which has been developed for logistics tracking within the framework of the ParsiFAl4.0 project [73]. **a** Electrical characterization of the UTCs embedded in CFP [17]. **b** CFP flexible package with two embedded UTCs while being released from the rigid carrier wafer. **c** The CFP fan-out package is assembled as an interposer in an LCP flexible motherboard with coarser interconnections. **d** The smart label, including the CFP interposer, is bent on a cylinder to demonstrate its mechanical flexibility

provides a strong-enough electromagnetic field that powers the nearby tag(s). Driven by Moore's law and the rise of artificial intelligence, more and more computation power is implemented locally in the sensor nodes. That's why, the next generation of RFID tags requires not only communication and memory blocks, but also signal processing cores, support for various communication protocols, integrated sensors and sensory peripherals for off-chip sensors [70, 71].

For the purpose of demonstrating the practical feasibility of the CFP embedding technology, various smart sensor systems have been developed. Utilizing the CFP as a standalone SoP, an e-skin for a bionic handling assistant has been implemented (cf. section 4.5), highlighting the key merits of HySiF, when the high-performance UTCs and the large-area passive and active components are complementary integrated into the CFP polymeric foil [14, 15].

Figure 2.8a, b show two UTCs, namely the Apollo microcontroller and the Near-Field Communication (NFC) IC (manufactured by the company em microelectronic [72]), which are embedded in the CFP. The flexible package is assembled on LCP flexible board as a fan-out interposer, alongside other digital sensors and communication ICs, as shown in Fig. 2.8c. This high-performance smart label has been developed for logistics tracking within the framework of the ParsiFAl4.0 project [73]. The off-chip/on-foil spiral inductor, which is presented in Sect. 3.4, is designed to be the flexible antenna of the embedded NFC IC.

References

1. Burghartz JN, Alavi G, Albrecht B, Deuble T, Elsobky M, Ferwana S, Harendt C, Mahsereci Y, Richter H, Yu Z (2019) Hybrid systems-in-foil-combining the merits of thin chips and of large-area electronics. IEEE J Electron Devices Soc 7:776–783
2. (2020) NXP semiconductors. Available https://www.nxp.com/products/security-and-authentication/security-controllers/ultra-thin-contactless-chip-module-mob10:MOB10
3. Infineon Technologies AG (2020). Available https://www.infineon.com/cms/de/about-infineon/press/market-news/2018/INFDSS201811-021.html
4. Xu S, Jayaraman A, Rogers J (2019) Skin sensors are the future of health care. Nature 571:319–321
5. IDTechEX (2020) Available http://www.idtechex.com/de/research-report/electronic-skin-patches-2019-2029/674
6. Ganasan JR (1999) Chip on chip (COC) and chip on board (COB) assembly on flex rigid printed circuit assemblies. In: 1999 Proceedings. 49th Electronic components and technology conference (Cat. No. 99CH36299), June 1999, pp 174–176
7. Würth Elektronik Group (2020) Available https://www.we-online.de/web/de/leiterplatten/produkte/3dstarrflexibleleiterplatten
8. Lu S, Chen W (2010) Reliability and flexibility of ultra-thin chip-on-flex (UTCOF) interconnects with anisotropic conductive adhesive (ACA) joints. IEEE Trans Adv Pack 33(3):702–712
9. Christiaens W, Bosman E, Vanfleteren J (2010) UTCP: A novel polyimide-based ultra-thin chip packaging technology. IEEE Trans Components Pack Technol 33(4):754–760
10. Hassan M, Schomburg C, Harendt C, Penteker E, Burghartz JN (2013) Assembly and embedding of ultra-thin chips in polymers. In: 2013 Eurpoean microelectronics packaging conference (EMPC), Sep 2013, pp 1–6
11. Alavi G (2019) Hybrid system-in-foil integration and interconnection technology based on adaptive layout technique. Ph.D. dissertation, University of Stuttgart
12. American Semiconductor Homepage (2020). Available https://www.americansemi.com/flex-ics.html
13. Cauwe M, Vandecasteele B, Gielen A, De Baets J, van den Brand J, Kusters R (2011) Technology development for a low-cost, roll-to-roll chip embedding solution based on pet foils. In: 18th European microelectronics packaging conference, Sep 2011, pp 1–6
14. Hassan M, Keck J, Klauk H, Kostelnik J, Mahsereci Y, Sailer S, Schreivogel A, Zaki T, Burghartz JN (2015) Combining organic and printed electronics in hybrid system in foil (HySiF) based smart skin for robotic applications. In: 2015 European microelectronics packaging conference (EMPC), Sep 2015, pp 1–6
15. Elsobky M, Mahsereci Y, Yu Z, Richter H, Burghartz JN, Keck J, Klauk H, Zschieschang U (2018) Ultra-thin smart electronic skin based on hybrid system-in-foil concept combining three flexible electronics technologies. Electron Lett 54(6):338–340
16. Sterken T, Vanfleteren J, Torfs T, de Beeck MO, Bossuyt F, Van Hoof C (2011) Ultra-thin chip package (UTCP) and stretchable circuit technologies for wearable ECG system. In: 2011 Annual international conference of the IEEE engineering in medicine and biology society, Aug 2011, pp 6886–6889

17. Elsobky M, Deuble T, Ferwana S, Albrecht B, Harendt C, Ottaviani A, Alomari M, Burghartz JN (2020) Characterization of on-foil sensors and ultra-thin chips for HySiF integration. IEEE Sens J 1–1
18. Khan S, Lorenzelli L, Dahiya RS (2015) Technologies for printing sensors and electronics over large flexible substrates: a review. IEEE Sens J 15(6):3164–3185
19. Heliatek Homepage (2020) Available https://www.heliatek.com
20. Yi-Bing C, Alex P, Fuzhi H, Yong P (2016) Print flexible solar cells. Springer Nature 11
21. Koden M (2017) OLED displays and lighting: flexible OLEDs. IEEE. Available https://ieeexplore.ieee.org/document/7827447
22. Wang C, Hwang D, Yu Z, Takei K, Park J, Chen T, Ma B, Javey A (2013) User-interactive electronic skin for instantaneous pressure visualization. Nat Mater 12:07
23. Huang Y, Chen J, Yin Z, Xiong Y (2011) Roll-to-roll processing of flexible heterogeneous electronics with low interfacial residual stress. IEEE Trans Components Pack Manuf Technol 1(9):1368–1377
24. Hassan M-U (2017). Ultra-thin chip embedding and interconnect technology for system-in-foil applications. Ph.D. dissertation, University of Stuttgart. Shaker Verlag
25. Alavi G, Sailer H, Albrecht B, Harendt C, Burghartz JN (2018) Adaptive layout technique for microhybrid integration of chip-film patch. IEEE Trans Components Pack Manuf Technol 8(5):802–810
26. Burghartz JN, Appel W, Harendt C, Rempp H, Richter H, Zimmermann M (2009) Ultra-thin chips and related applications, a new paradigm in silicon technology. In: 2009 Proceedings of the European solid state device research Cconference, Sep 2009, pp 29–36
27. (2020) Chip-flex technology: hybrid systems-in-foil. In: Elements in flexible and large-area electronics. Cambridge University Press
28. Lenihan T, Schaper L, Shi Y, Morcan G, Parkerson J (1996) Embedded thin film resistors, capacitors and inductors in flexible polyimide films. In: 1996 Proceedings 46th electronic components and technology conference, May 1996, pp 119–124
29. Elsobky M, Alavi G, Albrecht B, Deuble T, Harendt C, Richter H, Yu Z, Burghartz J (2018) Ultra-thin sensor systems integrating silicon chips with on-foil passive and active components. Proceedings 2:748
30. Cheng PL, Leung SYY, Law TW, Liu CK, Chong JIT, Lam DCC (2007) Quantitative analysis of resistance tolerance of polymer thick film printed resistors. IEEE Trans Components Pack Technol 30(2):269–274
31. Cook BS, Cooper JR, Tentzeris MM (2013) Multi-layer RF capacitors on flexible substrates utilizing inkjet printed dielectric polymers. IEEE Microw Wirel Components Lett 23(7):353–355
32. Burghartz JN (2011) Ultra-thin chip technology and applications. Springer, New York
33. Gupta S, Navaraj W, Lorenzelli L, Dahiya R (2018) Ultra-thin chips for high-performance flexible electronics. NPJ Flexible Electron 2
34. Burghartz JN (2013) You can't be too thin or too flexible, vol 50, Spectrum, IEEE. pp 38–61
35. Takyu S, Kurosawa T, Shimizu N, Harada S (2006) Novel wafer dicing and chip thinning technologies realizing high chip strength, vol 970, p 5
36. Hassan M, Angelopoulos EA, Rempp H, Endler S, Burghartz J-N (2010) Packaging challenges associated with warpage of ultra-thin chips. In: 3rd Electronics system integration technology conference ESTC, Sep 2010, pp 1–5
37. Fujisaki Y, Koga H, Nakajima Y, Nakata M, Tsuji H, Yamamoto T, Kurita T, Nogi M, Shimidzu N (2014) Transparent nanopaper-based flexible organic thin-film transistor array. Adv Funct Mater 24:03
38. Kraft U, Zaki T, Letzkus F, Burghartz J, Weber E, Murmann B, Klauk H (2018) Low-voltage, high-frequency organic transistors and unipolar and complementary ring oscillators on paper. Adv Electron Mater 5:11
39. Elsobky M, Alavi G, Zschieschang U, Aghamohammadi M, Hassan M-U, Klauk H, Harendt C, Richter H, Burghartz JN (2016) Application of BCB-on-Si based substrate in combining flexible organic electronics and high frequency antenna. In: International symposium on organic electronics: recent developments and challenges, 06 2016

40. Afsar Y, Moy T, Brady N, Wagner S, Sturm JC, Verma N (2017) 15.1 large-scale acquisition of large-area sensors using an array of frequency-hopping ZnO thin-film-transistor oscillators. In: 2017 IEEE international solid-state circuits conference (ISSCC), Feb 2017, pp 256–257
41. Tian B, Cohen-Karni T, Qing Q, Duan X, Xie P, Lieber C (2010) Three-dimensional, flexible nanoscale field-effect transistors as localized bioprobes. Science (New York, N.Y.) 329:830–834
42. Khodagholy D, Doublet T, Quilichini P, Gurfinkel M, Leleux P, Ghestem A, Ismailova E, Hervé T, Sanaur S, Bernard C, Malliaras G (2013) In vivo recordings of brain activity using organic transistors. Nat Commun 4:1575
43. Robinson JT, Pohlmeyer E, Gather MC, Kemere C, Kitching JE, Malliaras GG, Marblestone A, Shepard KL, Stieglitz T, Xie C (2019) Developing next-generation brain sensing technologies—A review. IEEE Sens J 19(22):10163–10175
44. Albrecht B, Alavi G, Elsobky M, Ferwana S, Passlack U, Harendt C, Burghartz JN (2018) Multi-chip patch in low stress polymer foils based on an adaptive layout for flexible sensor systems. In: 2018 7th Electronic system-integration technology conference (ESTC), Sep 2018, pp 1–5
45. Hua Q, Sun J, Liu H, Bao R, Yu R, Pan C, Wang Z (2018) Skin-inspired highly stretchable and conformable matrix networks for multifunctional sensing. Nat Commun 9:12
46. Niu S, Matsuhisa N, Beker L, Li J, Wang S, Wang J, Jiang Y, Yan X, Yun Y, Burnett W, Poon S, Tok J, Chen X, Bao Z (2019) A wireless body area sensor network based on stretchable passive tags. Nat Electron 2:1–8
47. Ghanbari MM, Tsai JM, Nirmalathas A, Muller R, Gambini S (2017) An energy-efficient miniaturized intracranial pressure monitoring system. IEEE J Solid-State Circ 52(3):720–734
48. Parker RL, Krinsky A (1963) Electrical resistance-strain characteristics of thin evaporated metal films. J Appl Phys 34(9):2700–2708
49. Baetens T, Pallecchi E, Thomy V, Arscott S (2018) Cracking effects in squashable and stretchable thin metal films on PDMS for flexible microsystems and electronics. Sci Rep 8:12
50. Widlund T, Yang S, Hsu Y-Y, Lu N (2014) Stretchability and compliance of freestanding serpentine-shaped ribbons. Int J Solids Structures 51:08
51. Hocheng H, Chen C-M (2014) Design, fabrication and failure analysis of stretchable electrical routings. In: Sensors (Basel, Switzerland) 14:11855–11877
52. Jeon S, Lim S, Trung TQ, Jung M, Lee N (2019) Flexible multimodal sensors for electronic skin: Principle, materials, device, array architecture, and data acquisition method. Proc IEEE 107(10):2065–2083
53. Jung M, Lee E, Kim D, Kim K, Yun C, Lee H, Kim H, Rhie K, Jeon S (2019) Amorphous FeZr metal for multi-functional sensor in electronic skin. NPJ Flexible Electron 3:8
54. Sun D-M, Timmermans M, Tian Y, Nasibulin A, Kauppinen E, Kishimoto S, Mizutani T, Ohno Y (2011) Flexible high-performance carbon nanotube integrated circuits. Nat Nanotechnol 6:156–161
55. Sun Y, Chang M, Meng L, Wan X, Gao H, Zhang Y, Zhao K, Sun Z, Li C, Liu S, Wang H, Liang J Chen Y (2019) Flexible organic photovoltaics based on water-processed silver nanowire electrodes. Nat Electron 11
56. Li W, Yang S, Shamim A (2019) Screen printing of silver nanowires: balancing conductivity with transparency while maintaining flexibility and stretchability. NPJ Flexible Electronics 3
57. Wacker N (2013) Characterization and compact modeling of complementary metal-oxide-semiconductor transistors for flexible electronics. Ph.D. dissertation, University of Stuttgart
58. Nguyen TT, Jeon S, Kim D-I, Trung TQ, Jang M, Hwang B-U, Byun K-E, Bae J, Lee E, Tok J, Bao Z, Lee N-E, Park J-J (2014) A flexible bimodal sensor array for simultaneous sensing of pressure and temperature. Adv Mater (Deerfield Beach, FL) 26:796–804
59. Kim K, Jung M, Kim B, Kim J, Shin K, Kwon O-S, Jeon S (2017) Low-voltage, high-sensitivity and high-reliability bimodal sensor array with fully inkjet-printed flexible conducting electrode for low power consumption electronic skin. Nano Energy 41:09

60. Ahmed S, Tahir FA, Shamim A, Cheema HM (2015) A compact Kapton-based inkjet-printed multiband antenna for flexible wireless devices. IEEE Antennas Wirel Propag Lett 14:1802–1805
61. Tang M, Shi T, Ziolkowski RW (2015) Flexible efficient quasi-yagi printed uniplanar antenna. IEEE Trans Antennas Propag 63(12):5343–5350
62. Simorangkir RBVB, Yang Y, Esselle KP, Zeb BA (2018) A method to realize robust flexible electronically tunable antennas using polymer-embedded conductive fabric. IEEE Trans Antennas Propag 66(1):50–58
63. Augustine R, Alves T, Zhadobov M, Poussot B, Sarrebourse T, Sauleau R, Thomas MK, Laheurte J (2010) SAR reduction of wearable antennas using polymeric ferrite sheets. In: Proceedings of the fourth European conference on antennas and propagation, April 2010, pp 1–3
64. Gao B, Yuen M, Ye T (2017) Flexible frequency selective metamaterials for microwave applications. Sci Rep 7:45108
65. García Núnez C, Manjakkal L, Dahiya R (2019) Energy autonomous electronic skin. NPJ Flexible Electron 3
66. Hu Y, Sun X (2014) Flexible rechargeable lithium ion batteries: advances and challenges in materials and process technologies. J Mater Chem A 2:10712–10738. Available http://dx.doi.org/10.1039/C4TA00716F
67. Cha H, Kim J, Lee Y, Cho J, Park M (2017) Issues and challenges facing flexible lithium-ion batteries for practical application. Small 14:1702989
68. Manjakkal L, Navaraj W, García Núñez C, Dahiya R (2019) Graphene–graphite polyurethane composite based high–energy density flexible supercapacitors. Adv Sci 6
69. Ozer E, Kufel J, Biggs J, Brown G, Myers J, Rana A, Sou A, Ramsdale C (2019) Bespoke machine learning processor development framework on flexible substrates. in: 2019 IEEE international conference on flexible and printable sensors and systems (FLEPS), July 2019, pp 1–3
70. Reinisch H, Wiessflecker M, Gruber S, Unterassinger H, Hofer G, Klamminger M, Pribyl W, Holweg G (2011) A multifrequency passive sensing tag with on-chip temperature sensor and off-chip sensor interface using EPC HF and UHF RFID technology. IEEE J Solid-State Circ 46(12):3075–3088
71. Deepu CJ, Zhang X, Liew W, Wong DLT, Lian Y (2014) An ECG-on-chip with 535 nW/channel integrated lossless data compressor for wireless sensors. IEEE J Solid-State Circ 49(11):2435–2448
72. Em Microelectronic Homepage (2019) Available https://www.emmicroelectronic.com/
73. Parsifal4.0 Project Homepage (2020) Available https://parsifal.ims-chips.de/html/projekt.html

Chapter 3
Off-Chip/On-Foil Passive and Active Components

HySiF concept gains its strength from mixing and matching different flexible electronic technology directions in a way that optimizes the utilization of its integrated components. Towards the realization of a mature and reliable HySiF based on the CFP process, different off-chip/on-foil passive and active components are implemented. In this chapter, six case studies are designed to highlight key HySiF challenges and present a practical methodology to address these challenges. The chapter starts with introducing three ultra-thin environmental sensors, namely one temperature and two relative humidity sensors. Here, the CFP compatibility with different sensing materials is investigated. Furthermore, on-foil spiral inductors are implemented to study the challenges associated with integrating wireless communication and energy harvesting functionality into HySiF. Finally, a low-voltage digital library for Organic Thin-Film Transistors (OTFTs) is characterized on the CFP flexible substrate. The digital library is then used to implement an addressing circuit for distributed large-area sensor arrays.

3.1 On-Foil Resistive Temperature Detector

Figure 3.1 shows a conceptual schematic of one HySiF variant in which an off-chip/on-foil temperature sensor and UTC are integrated into the CFP flexible package [1]. In general, small and constant thermal contact resistance is required between the temperature sensing element and the surface or object that its temperature is measured (cf. thermal interface A in Fig. 3.1). Modern silicon-based integrated temperature sensors are compact, power-efficient and highly accurate [2]. However, the thermal insulation of the die plastic packaging, including the CFP, increases the thermal resistance between the sensing chip and the package surface, in addition to non-linearities in the sensor response. Albeit, on-chip temperature sensors can be used to monitor the chip self-heating and in-package thermal behavior in

M. Elsobky, *Ultra-Thin Sensors and Data Conversion Techniques for Hybrid System-in-Foil*, Recognizing Outstanding Ph.D. Research,
https://doi.org/10.1007/978-3-030-97726-9_3

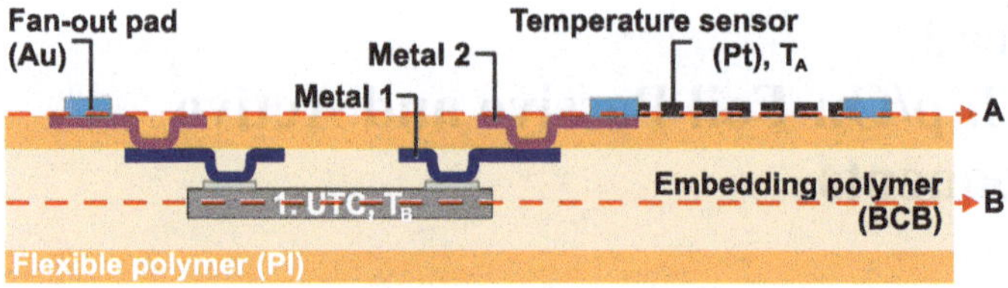

Fig. 3.1 Cross-section of one variant of the CFP process in which embedded UTC and off-chip/on-foil temperature sensor are integrated into/on a flexible polymeric substrate [1]

high-frequency communication or high-power switching applications (cf. thermal interface B in Fig. 3.1) [3, 4].

To resolve this issue, off-chip/on-foil thin-film temperature sensors are implemented in this work. The typical ultra-thin form factor and low thermal mass (i.e. small form factor) of thin-film Resistive Temperature Detectors (RTDs) allow them to draw negligible heat from the measured object or surface. Even when the thin-film RTD is encapsulated and its area increases much larger than that of the on-chip sensors, the thermal resistance between the thin-film RTD and the package surface would still be small. These thermal characteristics when combined with the thin and adaptive form factor of the CFP package, enables the realization of precision bendable smart temperature sensors.

Different flexible temperature sensing elements have been reported (e.g. flexible amorphous silicon [5], carbon black and nickel oxide mixture [6] and printed graphene oxide [7]). Here, thin-film platinum (Pt) RTDs are chosen due to their well-defined temperature response and inherent compatibility with standard microelectronics fabrication processes that are used during UTC embedding in the CFP process.

Sensor Design

As shown in Fig. 3.2, the RTDs use the serpentine and horseshoe layouts. The former is used when a compact sensor area is needed and the latter is used when stress compensation, mechanical compliance and stretchability are crucial. The serpentine metal linewidth and spacing are both about 10 μm, which is limited by the photolithography processing step. Large contact pads ($250\,\mu m \times 250\,\mu\text{m}$) are used to allow enough space for 4-point probing during the sensor characterization.

For the choice of the optimal horseshoe shape under different material and dimension constraints, the decision criteria from [8] is followed. Figure 3.2d shows the different design parameters of the horseshoe layout, where R is the radius of the arc median line, w is the linewidth, α is the arc angle, L is the arm length and S is the end-to-end distance of the unit horseshoe cell. Horseshoe layouts with narrower linewidth w are more compliant and stretchable [8], which is why the minimum linewidth of 10 μm is chosen. A compact sensor layout is also targeted, hence the arm length L is minimized. Three unknowns α, w/R and L/R are solved as in [8], under several constraints (e.g. the distance between the two nearest horseshoe ribbons,

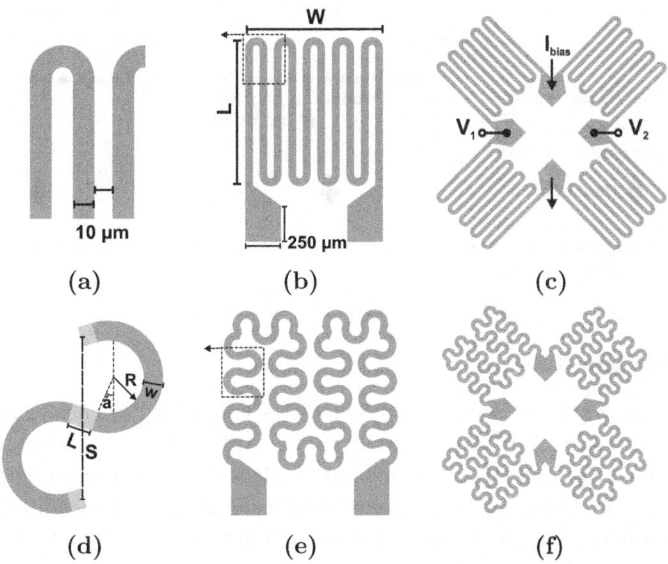

Fig. 3.2 Temperature sensor design. **a–c** Serpentine and **d–f** horseshoe RTD layouts where minimum linewidth and spacing are 10 μm. Wheatstone bridge configuration using (**c**) serpentine and (**e**) horseshoe unit RTDs in which the resistance mismatch R_{offset}, $\Omega = (V_1 - V_2)/I_{Bias}$ is extracted

surface area and minimizing the effective applied strain). The parameters α, R, L and S are chosen to be about 30 °, 50 μm, 5 μm and 140 μm, respectively.

Figure 3.2c, f show the RTDs when used as unit elements in Wheatstone bridge configuration. The purpose of this design is to directly quantify the resistance mismatch, which results from fabrication non-idealities in addition to testing possible fully differential instrumentation. Note, that evaporated thin-film metal lines show elevated gauge factors than those for bulk metals (gauge factors upto 20 for thin-film vs. 2 for bulk metals) [9, 10], which calls for system-level cancellation of common-mode disturbance, such as mechanical stress, by employing differential signaling. Here, the resistance mismatch is extracted using R_{offset}, $\Omega = (V_1 - V_2)/I_{Bias}$, where V_1 and V_2 are the bridge outputs and I_{Bias} is the bridge constant biasing current.

Sensor Fabrication

Figure. 3.3 shows three different fabrication setups that are used to characterize the on-foil RTDs [1]. In the first setup and using a 150-mm silicon wafer, 50-nm Pt is evaporated on top of 10-nm titanium (Ti) thin-film layer, which acts as adhesion promoter. For the second and third setups, the same sensor stack is evaporated on a layer of 1-μm-thick BCB, which is spin-coated on a silicon wafer. A standard lift-off process is performed for the three setups. For the third setup, an additional layer of 1-μm-thick BCB is used to encapsulate the Ti/Pt sensor stack and the BCB is finally etched to open larger contact pads.

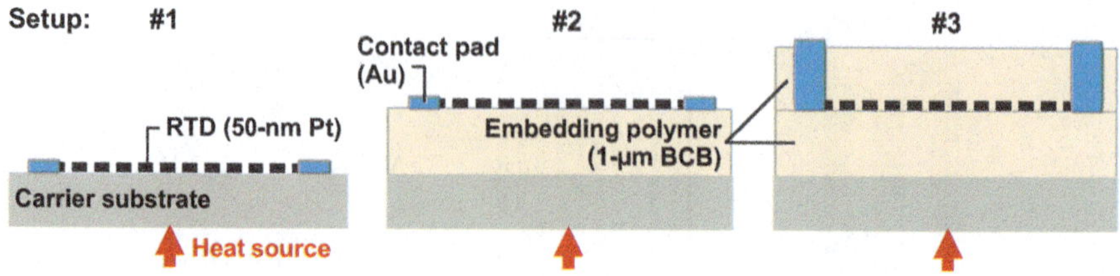

Fig. 3.3 Cross-section of three setups used to electrically and thermally characterize the thin-film Pt RTD [1]. Not drawn to scale

Sensor Characterization

The RTD wafers are placed on a thermo-chuck, which is used as a heat source and its temperature is varied starting from 40 till 120 °C in 20°C steps. At room temperature (about 20 °C), an extra data point is measured. The substrate surface temperature (in setup #1 SiO_2, in #2 and #3 BCB) is monitored using a reference thin-film Pt100 foil sensor from which the worst-case inaccuracy in the quasi steady state temperature measurement equals to ±0.75 °C.

Different RTDs with various surface areas are measured and the nominal resistance at room temperature (about 20 °C) ranges from 80 to 50 kΩ for surface areas from 0.06 to 2.5 mm^2. Figure 3.4a shows the static characterization of three RTDs with similar dimensions and horseshoe layouts but from different setups (cf. Fig. 3.3) [1]. The variation in the base resistance value originates from the different substrate material and surface topography. The extracted sensitivities ($\Delta R/R_{20\,^\circ C}$) of 18 different sensors are plotted in Fig. 3.4b, reflecting temperature coefficients ranging from 2 to 3 mΩ/mΩ/°C depending on the substrate composition.

Although the surface temperature in setup #2 is less than that in setup #1 due to the thermal barrier introduced by the BCB layer, setup #2 clearly exhibits better sensor sensitivity. This is attributed to a higher quality of the Pt metal film when fabricated on the smoother BCB surface [11] against the silicon oxide surface [12] of setup #1. Sensors in setup #3 show the highest effective sensitivity due to the presence of an additional BCB thin thermal barrier layer between the RTD and air, thus lowering the heat loss to ambient air by convection.

Figure 3.4c shows the bridge offset resistance value R_{offset} for the horseshoe- and serpentine-based Wheatstone bridges. Although both bridges have the same surface area (4 × 2.5 mm^2), the horseshoe-based Wheatstone bridge showed superior matching to that of the serpentine-based bridge. This is regarded to better randomization of the fabrication process non-idealities for the horseshoe layout. For instance, the laser direct writer scans the wafer in one direction only, which leads to either better or less matching between the structures fabricated parallel or perpendicular to each other, respectively.

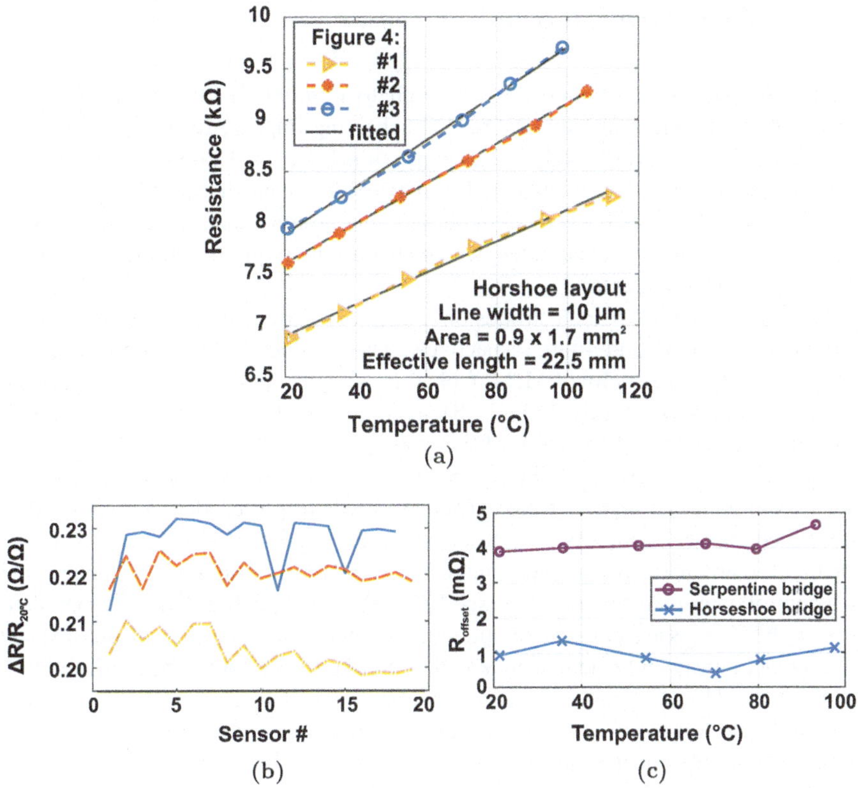

Fig. 3.4 **a**Measured resistance of three RTDs with similar dimensions from three different setups. **b** Measured sensitivity of multiple RTDs from the three different setups shown in Fig. 3.3. **c** Bridge offset resistance R_{offset} plotted for Wheatstone bridge RTDs designed using serpentine and horseshoe layouts

3.2 Ultra-Thin Humidity Sensors

Capacitive polymeric Relative Humidity (RH) sensors are currently dominating the market [13, 14]. Other developing resistive or electrochemical-based RH sensors, such as graphene oxide (GO) [15, 16], nanosheets [17, 18] and high-surface-to-volume ratio materials, such as carbon nanotubes and nanowires [19, 20], have the potential to achieve faster response time and higher sensitivities. Here, two variants of RH sensors are presented, namely electrochemical and capacitive sensors, which have proven to be compatible with the CFP flexible package [21].

3.2.1 Electrochemical Humidity Sensor

Previously, nanosheets composed of phosphatoantimonic acid $H_3Sb_3P_2O_{14}$ were spin-coated on rigid glass substrates [17] achieving huge sensitivity (5 orders of magnitude from 0 to 100% RH) and response time of $\approx$ 2 s. Upon moisture absorption, water molecules get intercalated in between the nanosheets, which increases the film thickness. This leads to an increase in the proton conductance and a simultaneous shift in the refractive index, which enables the implementation of electrical as well as optical readouts.

Here, sputtered AlSiCu electrodes on BCB/PI substrate define the sensor length and width as 0.5mm and 4mm, respectively. These values were chosen to keep the sensor base resistance below 1 MΩ (cf. [17]) for simpler sensor readout. Afterwards, the same nanosheets ($H_3Sb_3P_2O_{14}$ [17]) are spin-coated on the BCB/PI polymeric substrate at a speed of 2000rpm for 90seconds.

For sensor characterization, the electrochemical sensor is assembled using double-sided foam adhesive on a flexible carrier PCB, where pin headers are soldered to predefined copper contact pads. The electrical connections from the sensor to the carrier PCB are made using silver glue (EPO-TEK H20E). The sensor is then placed in a controlled climate chamber (Vötsch VCL 0010) and the RH is varied from 45 to 80% with 5% steps at a constant temperature of 30 °C. The chamber sensors are used as reference sensors with an inaccuracy of 0.58 K and 1.7% for temperature and RH, respectively. The RH sensor is then connected to an LCR meter (HP4284A, input amplitude and frequency of 300 mVpp and 800 Hz) using 4-point Kelvin connection for electrical impedance characterization. Note that a higher amplitude of the AC excitation signal can permanently ionize, and, thus, destroy the nanosheets.

Figure 3.5 shows the characterization results of the electrochemical RH sensor. The relative conductance change $\Delta G/G_{RH=50\%}$ is plotted against the RH data from the reference sensor in the RH range from 45 to 80%. A linear fit is also plotted with a coefficient of determination ($R^2 = 0.9724$). The extrapolated fit indicates an expected maximum $\Delta G/G_{RH=50\%}$ of about $\pm$368% for the full RH range. The measured sensor performance matches well with the previous realization of the same sensing material on rigid glass. This indicates that the BCB surface of the CFP package is smooth enough on the nanoscale level for the proper fabrication and operation of the nanosheets. The inset of Fig. 3.5a shows the change in the film optical thickness (surface color from red to green) at RH of 45 and 100%. Figure 3.5b shows the Nyquist plot of the electrochemical sensor at RH of about 60%. The measurement results indicate that the sensor electrical behavior can be modeled using Randles equivalent circuit [22] that is shown in the inset of Fig. 3.5b. The extracted proton conductivity is about 15 mS/cm at RH of about 55%.

Although the implemented electrochemical sensor provides huge sensitivity and optical readout, it still requires AC excitation and complex impedance readout techniques for integration into HySiF. Due to the reactive electrochemical nature of the nanosheets, a protective vent can be assembled onto the sensing active region in order to reduce condensation and contamination [23].

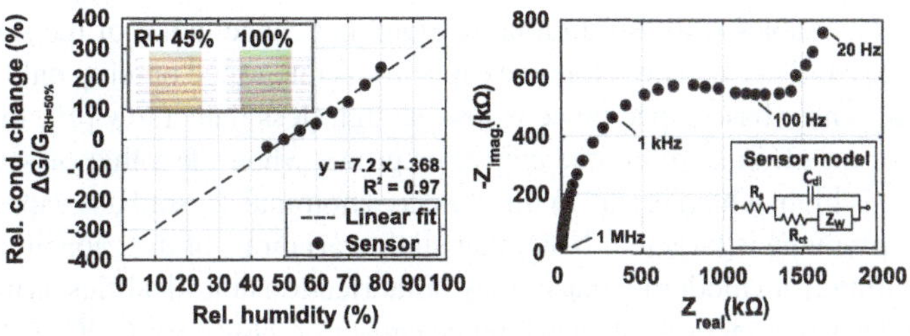

(a) (b)

Fig. 3.5 **a** Measurement results of the electrochemical sensor relative conductance change $\Delta G/G_{RH=50\%}$ against relative humidity. A linear fit is also plotted with a coefficient of determination (R^2 = 0.9724). The inset of this figure shows the change in the film optical thickness (surface color from reddish to green) at RH of 45 and 100%. **b** Nyquist plot of the electrochemical sensor at RH of about 60%. The measurement results indicate that the sensor electrical behavior can be modeled using Randles equivalent circuit [22], which is shown in the figure inset. The model consists of the electrolyte resistance R_s, charge transfer resistance R_{ct}, specific electrochemical element of diffusion Z_w and double-layer capacitance C_{dl}

3.2.2 Capacitive Humidity Sensor

Driven by the well-established plastics industry, polymers have been widely employed as dielectric material in integrated interdigitated and parallel plate capacitive RH sensors [13, 14, 24–26]. They currently dominate the RH sensor market due to their linear response, CMOS-compatibility, stable operation over long periods and potential high power efficiency due to the fact that no static power is consumed to operate the sensors [24, 27–30]. Thanks to the automotive and consumer electronics applications, environmental monitoring sensors, RH sensors in particular, are directed towards miniaturization and compactness as the polymer sensing material is typically spin-coated on the top metal layer of the CMOS readout chip. Consequently, wire-bonding between the sensor and readout circuit is eliminated. A one-chip solution is achieved where short circuit interconnections result in low parasitic capacitance, which enables compact, ultra-low power and precision sensor readout [29]. However, this technique is not economic for flexible and large-area electronic applications (area of more than tens of mm^2, e.g. touchless motion tracking), that's why an inherently CFP-compatible off-chip/on-foil polymeric capacitive RH sensor for HySiF integration is developed in this work.

Sensor Design

The RH sensor implemented in this work uses the variations in the dielectric constant ε_r of PI material (Durimide manufactured by the company Fujifilm [31]) upon moisture absorption and desorption [21]. The sensing material is inherently compatible with the CFP UTC-embedding process as it has been used as a reinforcing polymer in one variant of the CFP flexible package [32].

Besides, most integrated capacitance-to-digital converters (CDCs) implementations use an insensitive and stable reference capacitor to construct a half or full bridge, which enables fully differential readout and cancellation of the sensor base capacitance [24, 27–29]. The reference capacitor is integrated on-chip only when the value of the sensor base capacitance is low enough (less than a few pF). In contrast, the need for an off-chip reference capacitor grows when the value of sensor base capacitance is high, which is true in this case as a nominal sensor base capacitance in the tens of pF range is targeted. Note, that slight variations of the capacitor nominal value, resulting from modeling inaccuracy or fabrication non-idealities, is mostly tolerated in CDCs that incorporate a reference capacitor. However, CDCs that directly digitize the sensor base capacitance along with the RH capacitive variations suffer from degraded input dynamic ranges upon variations in the value of the sensor base capacitance.

The capacitance ratio of the wet (RH=100%) and dry (RH=0%) PI is given in first approximation by the following relation [33]:

$$\frac{C_{\text{wet}}}{C_{\text{dry}}} = \frac{\varepsilon_{\text{wet}}}{\varepsilon_{\text{dry}}}. \tag{3.1}$$

Using the empirical Looyenga formula [34], we can calculate ε_{wet} for different fractional volume γ of absorbed water in our PI film as follows [33]:

$$\varepsilon_{\text{wet}} = \left\{\gamma \left(\varepsilon_{\text{water}}^{1/3} - \varepsilon_{\text{dry}}^{1/3}\right) + \varepsilon_{\text{dry}}^{1/3}\right\}^3, \tag{3.2}$$

where $\varepsilon_{\text{water}} = 78$. Given that the capacitance value of the fringe capacitor is a linear function of its dielectric constant ε_r to the first approximation [24, 33, 35] and that for our PI, ε_r changes from 3.5 to 3.8 when RH changes from 4 to 50% (curing temperature of 350 °C for 2 h, cf. manufacturer's datasheet [31]), an ideal relative sensitivity $S = \Delta C/C_{RH=50\%}$ of about ±8% is expected.

To contain all the fringing electric field lines of the planar capacitor, the PI thickness should be at least equal to the sum of the finger width and spacing [33]. Here, the minimum resolution is limited by the lithography to about 10μm that in turn requires a PI thickness of 20 μm. Consequently, more than 6 spin-coating and curing cycles are required to achieve this PI thickness. As proven in [36], the thickness of PI must be minimized for faster sensor response. In this work, we have used two PI thicknesses, namely 3 and 1 μm, in order to achieve fast response time and a simpler fabrication process.

Using the analytical model presented in [35], a capacitance per unit area value of 1.16 pF/mm^2 is calculated for a fringe capacitor with finger width and spacing of 10 μm, PI layer thickness of 3 μm and ε_r of 3.8. Finite Element Method (FEM) is also used to calculate the nominal value of the sensor base capacitance where the low-frequency solver of the CST studio suite is used. A capacitance per unit area value of 1.57 pF/mm^2 is simulated for the same previously mentioned dimensions and material properties.

Measurements, using LCR meter and 4-point connection, show a close value of 1.28 and 1.02 pF/mm^2 for sensors (finger width and spacing of 10 μm) with 3- and

1-µm-thick PI sensing layers, measured at RH of 55% and temperature of 25 °C. Here, the sense and reference capacitors use the interdigitated electrode design for achieving a simple planar fringe capacitor. Different finger width and spacing are designed ranging from 5 to 20 µm in a fixed area of 5 × 5 mm^2

Sensor Fabrication

Figure 3.6 shows the cross-section of the BCB/PI flexible substrate on which the RH sensor and two reference capacitor variants are fabricated. Following the CFP process [32], the substrate is an alternate stack of BCB and PI layers spin-coated on a 150-mm carrier wafer. The metal electrodes are fabricated using sputtered AlSiCu that is structured using standard lithography step (laser direct writing, VPG400 HIMT, followed by dry etching). This optical lithography step, along with other process non-idealities, such as the BCB/PI surface topography, determines the minimum electrode spacing and consequently the maximum capacitance per unit area.

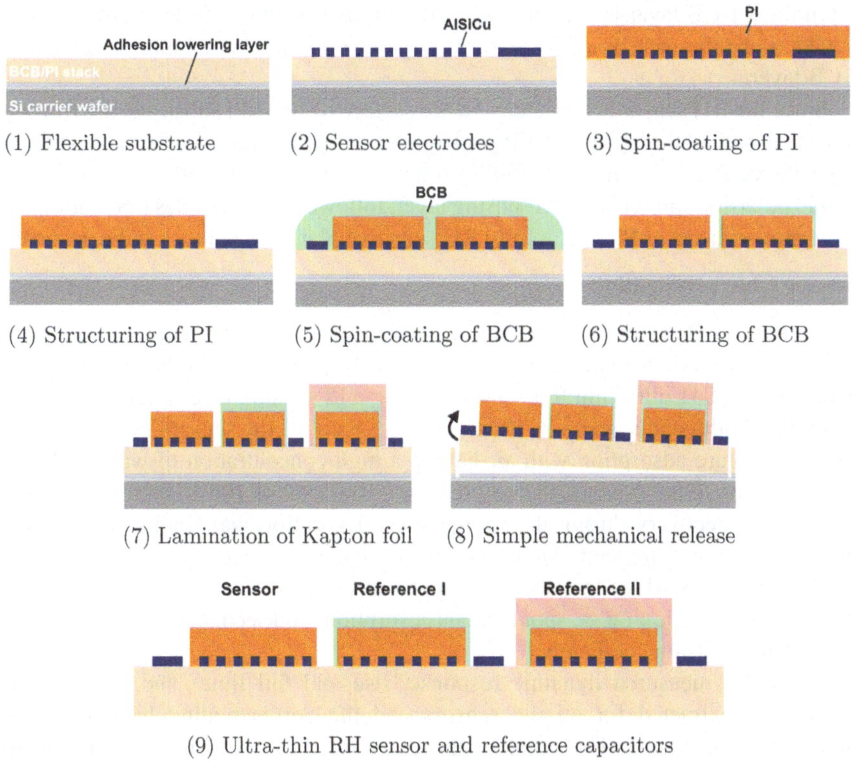

Fig. 3.6 Fabrication process steps for the ultra-thin RH sensor, reference I and reference II. The sensing material is PI, which is isolated using BCB to fabricate reference I and an additional Kapton lamination to fabricate reference II. The foil is detached (cf. Fig. 3.7) from the carrier wafer for characterization

(a) (b) (c)

Fig. 3.7 **a** RH sensors on 150-mm carrier wafer. Illustration of the RH sensor **b** during and **c** after the simple mechanical release or detachment from the carrier wafer

The PI sensing material is then spin-coated at 1500 rpm and cured at a temperature of 250 °C achieving a PI thickness of about 3 µm, which is etched down to achieve a PI thickness of 1 µm for a faster sensor response time. The fabrication of references I and II in Fig. 3.6 follows the exact processing steps of the RH sensor and an additional 1-µm thick BCB layer is spin-coated on the PI layer to shield it from any moisture uptake. For reference II, a 25-µm-thick Kapton foil is laminated on top of the thin BCB layer.

Since the RH sensor is fabricated on the surface of BCB/PI substrate, the fringing field that passes through the CFP package causes a humidity-insensitive parasitic capacitance in parallel to the humidity sensitive capacitor. An alternative approach would be to first spin-coat the PI sensing layer, followed by electrodes fabrication and then another layer of PI, thus a fully RH sensitive planar capacitor can be achieved.

Sensor Characterization

In this section, the dynamic behavior of the RH sensor is discussed, while the static characterization is presented later in section 4.1. Figure 3.8 shows the dynamic characterization of 41 ultra-thin RH sensors having a PI sensing layer thickness of 1µm and finger width and spacing of 10µm. Figure 3.8a plots the sensors' response upon forced moisture adsorption with air having a high concentration of water vapor in order to saturate the sensors, thereby reaching RH level of 100%. Figure 3.8b plots the sensors' recovery during the spontaneous desorption that occurs shortly after the air flow is discontinued. An asymmetry is observed between the adsorption and desorption events which traces primary to the different diffusion triggering mechanisms (forced adsorption vs. spontaneous desorption) and secondary to the moisture-dependent diffusion coefficient [36].

Using the measured dynamic response, rise and fall times and relative sensitivity are extracted for all the sensors and the corresponding histograms are plotted in Fig. 3.8c–e. The relative sensitivity (cf. Fig. 3.8d) is calculated using $(C_{max} - C_{min})/C_{min}$, where C_{max} and C_{min} are the maximum and minimum capacitance values at 100% RH and 55% RH, respectively. The average rise and fall times and relative sensitivity are 54, 750 ms and ±21.85%, respectively. Furthermore, the measured relative sensitivity is about 2.73x more than the analytically predicted relative sensitivity (±8%) in spite of using thinner PI sensing layers. Using lower

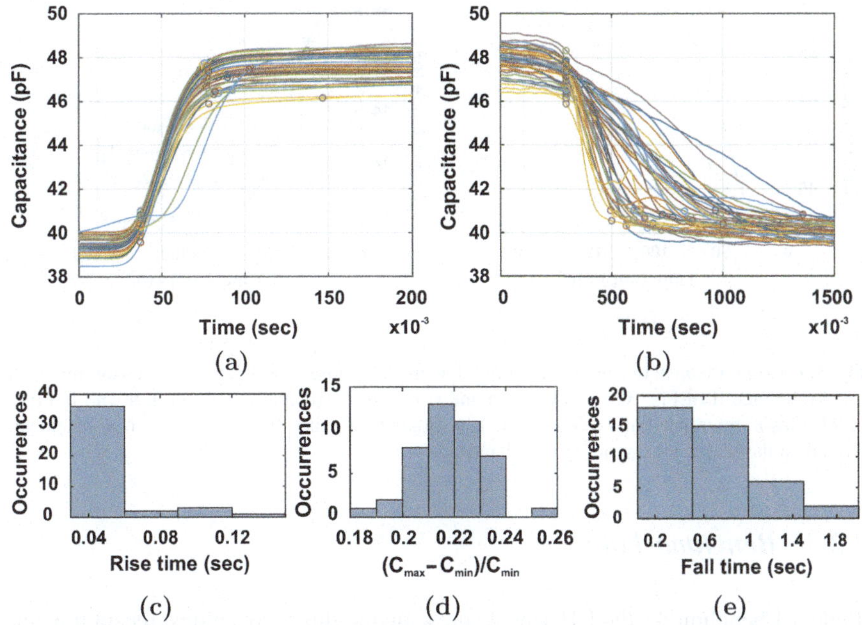

Fig. 3.8 Dynamic characterization of RH sensors where 41 sensors are measured by **a** forced adsorption of water vapor, thus saturating the sensor dielectric with 100% RH. Rise times are extracted and a histogram is plotted in (**c**) with mean time of 54 ms. **b** Sensors' recovery by spontaneous desorption. **d** Histogram of the relative sensitivity (calculated using $(C_{max} - C_{min})/C_{min}$), where C_{max} and C_{min} are the maximum and minimum capacitance values at 100% RH and 55% RH, respectively. Fall times are extracted and a histogram is plotted in (**e**) with mean time of 750 ms

curing temperature than that mentioned in the PI datasheet, might be a reason for the improved sensitivity.

Figure 3.9 compares the dynamic response of the RH sensor with reference I and II, where Fig. 3.9a, b plots the response and recovery times. By extracting the relative sensitivity of reference I ($\pm 2.71\%$), the additional 1-µm-thick BCB layer desensitized the sense capacitor by a factor of $8\times$ achieving a good reference capacitor while maintaining the same value of the base capacitance. A thicker BCB layer could completely desensitize the reference capacitor. The 25-µm-thick Kapton foil achieved complete desensitization for reference II, with a slightly higher capacitance value. Minor level variations in the measured response of reference II are observed that are triggered by the adsorption and desorption events, which are related to variations in temperature, pressure and contact resistance.

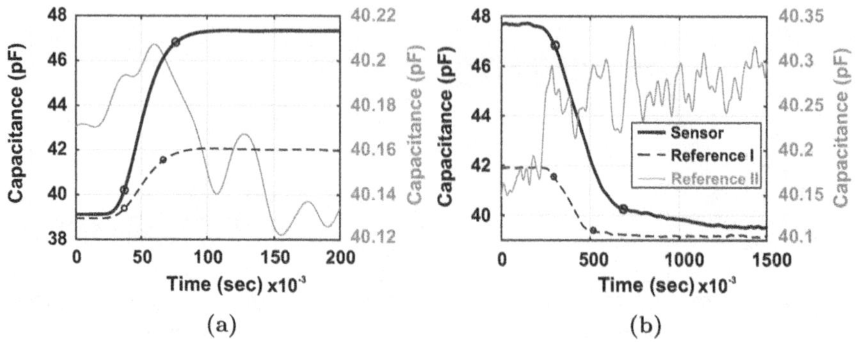

Fig. 3.9 Comparison between the measured dynamic response of RH capacitive sensor, reference I and reference II of Fig. 3.6. **a** Adsoption and rise time, **b** spontaneous desorption and fall time of RH sensor, reference I and reference II. The value of reference II capacitance is near-stable and plotted on the scaled y-axis on the right-hand side

3.2.3 Benchmarking

Table 3.1 benchmarks the RH sensors presented in this work against recent thin and ultra-thin RH sensors. Different sensing materials, with thicknesses ranging from few tens of nm to less than a mm, have been used, such as ceramics materials (e.g. Al_2O_3, TiO_2 and SiO_2) [20], polymers (e.g. Polyimides, Poly(ethyleneterephthalate) (PETT), Poly(methyl methacrylate) (PMMA) and Kapton) [25, 26] and inorganic materials (e.g. graphene oxide (GO), tungsten-based 2D semiconductors WO_3/WS_2/WS_3, 1D materials: nanofibers, nanowires, nanotubes) [15, 16, 19, 38, 40–42].

A wide range of RH detection techniques has been employed including those techniques, which are based on changes in the dielectric constant (capacitive), proton or ionic conduction (electrochemical, resistive or capacitive) and refractive index (optical). Sensor sensitivity or dynamic range vs. response time is typically optimized according to the target application. Few RH sensors have achieved response and recovery times in the ms range (e.g. GO-based [39], optical sensors [44] and thin-film polymer-based [45]) as most of the reported sensors have shown few tens of seconds response time.

3.3 Printed Strain Gauge

In this section, a resistive strain gauge, which is printed using silver-ink, is designed and characterized. In the next chapter (section 4.2), the presented strain gauge is combined with a readout UTC demonstrating a HySiF, which has been developed for robotic e-skin applications.

Table 3.1 Ultra-thin RH sensors performance summary and comparison against the state-of-the-art

Reference	Transduction	Sensing material	Thickness	Fabrication	Substrate	Electrode material	Sensitivity	Response time (s)	Recovery Time (s)	Measured range (%)
This work	EIS, R	nanosheets $H_3Sb_3P_2O_{14}$	0.2	Spin- coating	BCB/PI	AlSiCu/Au	±369%	2	2	45-80
This work	C	PI	1-3	Spin-coating	BCB/PI	AlSiCu/Au	±21%-90%	0.05-1	0.7-2.5	45-80
[26]	C	PI	4.6	Spin-coating	Glass	Au/Cu	15.2fF/%RH	18	31	5-85
[25]	C	Kapton	75	Inkjet Printing	Kapton	Ag-ink	0.5fF / %RH.mm^2	350		20-90
[37]	C	PEG/PEDOT:PSS	250	Spin-coating	PDMS	PEDOT:PSS	0.0085/%RH	2		10-90
[18]	R	PANI/PEDOT:PSS	100	Inkjet Printing	Paper	PEDOT:PSS	200%	420		16-98
[16]	EIS, R	GO	12	Spray-coating	Si/SiO_2	Au/Ti	4.7 M-0.6 kΩ	0.5		25-95
[15]	R	GO	0.0025	Drop-casting	Si/SiO_2	Au	1200x	5		25-88
[38]	C	GO	-	Drop-casting	Si/SiO_2	Au	37800%	10.5	40	15-95
[39]	EIS, R	GO	1-0.015	Spray-casting	PEN	Ag	$10_{-7} - 10_{-6}$	0.02	0.03	30-80
[40]	R	WS_2	0.0021+55	W +Sulfur-ization	PDMS	Graphene	2357x	5	6	20-90
[41]	EIS, R	WS_3	171	Spin-coating	Ceramic	Ag-Pd	277%	4	63	11-95
[42]	R	BP/Graphene	-	Electro-spraying	Si/SiO_2	Au	43%	9		15-70
[19]	R	nanofibers	0.02	Drop-coating	Glass	Au	45000x	2.2	1.05	5-80
[43]	EIS,C	IESM	19	Natural	IESM	Ag-ink	5 – 0.5 MΩ32 – 60pF	2		0-90
[44]	Optical	Nafion	1	Drop-coating	PDMS	-	1.4%/%RH	0.21	0.25	30-100

IESM: Inner Egg Shell Membrane, BP: Black Phosphorus, EIS: Electrochemical Impedance Spectroscopy, PDMS: Polydimethylsiloxane, PEN: Polyethylene naphthalate

Sensor Design

Generally, a wire's electrical resistance R is related to its length L and cross-section area A using the following relation:

$$R = \frac{\rho \cdot L}{A}, \tag{3.3}$$

where ρ is the material resistivity. As the wire stretches, its length increases and its cross-section area decreases by a certain ratio called Poisson ratio υ. The relative change in the electrical resistance is related to the mechanical strain ϵ by the gauge factor (GF) in the following relation:

$$GF = \frac{\Delta R/R}{\epsilon} = \frac{\Delta\rho/\rho}{\epsilon} + 1 + 2\upsilon, \tag{3.4}$$

where ΔR is the change in the electrical resistance. The bending strain at a distance y from the neutral surface of a bending plate is given by the following relation:

$$\epsilon = \frac{y}{r}, \tag{3.5}$$

where r is the bending radius of the neutral surface and 1/r is the bending curvature. We conclude from the relations (3.3)–(3.5) that bending the strain gauge with a certain radius results in a mechanical strain on its surface, which affects the geometric dimensions (macroscopic) and electrical resistivity (microscopic) resulting in variations of the strain gauge electrical resistance. Furthermore, the microscopic coefficient of piezoresistivity in metals is very small and can be neglected in most cases (but can not be neglected in thin-film metal [9, 10]) leaving only the macroscopic geometry variation factor to be evaluated.

The Wheatstone bridge circuit is a network of four series-parallel resistors as shown in Fig. 3.10a. It has two output terminals V_1 and V_2, which have equal voltage levels when the bridge is balanced. The bridge is balanced when the ratio of the two resistors in one branch is equal to the ratio of the two resistors in the other branch.

In order to achieve high bridge sensitivity and linear operation, full bridge topology is adopted in which two strain-sensitive resistors R_V, are printed vertically (cf. Fig. 3.10b) and the other two, ideally strain insensitive, resistors R_H are printed horizontally in the direction perpendicular to the bending strain. Equation (3.6) relates the bridge individual resistance values to the voltage difference between the output terminals V_{out}, given constant biasing current I_B:

$$V_{out} = V_1 - V_2 = I_B \cdot \frac{R_{V,1}R_{V,2} - R_{H,1}R_{H,2}}{R_{V,1} + R_{V,2} + R_{H,1} + R_{H,2}}. \tag{3.6}$$

It is important to note that the stability of the biasing current is crucial for the sensor readout accuracy, as any change in the biasing current is misinterpreted by

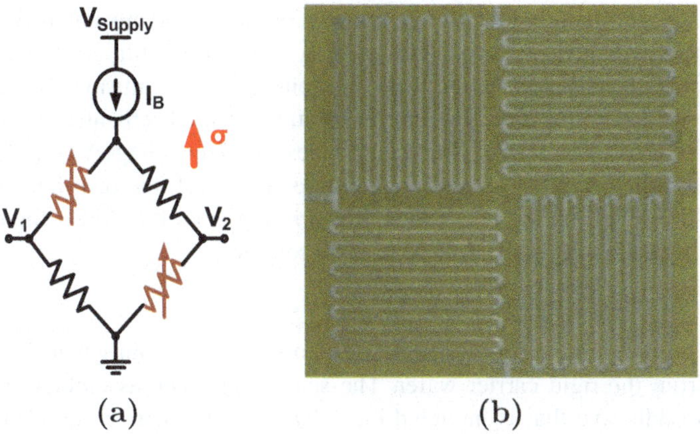

Fig. 3.10 **a** Wheatstone bridge circuit, which has four resistors arrange in a diamond shape, biased using constant current source. **b** Photograph of serpentine-shaped strain gauge printed using silver ink on BCB/PI substrate. The linewidth is 20μm, the total dimension of the bridge circuit is 5 x 5mm^2 and nominal resistance of 800Ω. The direction of expected bending (i.e. mechanical stress σ) of the robotic gripper is indicated

the bridge circuit as a change in the resistance of the active elements. The linear transfer characteristics of the bridge output voltage with the change in resistance ΔR can be shown using $R_V + \Delta R_V = R_{V,1} = R_{V,2}$ and assuming that the strain-insensitive resistors are equal ($R_H = R_{H,1} = R_{H,2}$) and using Eq. (3.6), this results in the following relation:

$$V_{out} = \frac{I_B}{2} \cdot \Delta R_V + V_{offset}, \tag{3.7}$$

where the offset voltage V_{offset} corresponds to the mismatch between the two printing directions and is given by $V_{offset} = \frac{I_B}{2}(R_V - R_H)$. Using Eqs. (3.4), (3.5) and (3.7) the end-to-end relation linking the bridge output voltage to the bending curvature is as follows:

$$V_{out} = \frac{I_B}{2} \cdot GF \cdot R_V \cdot \frac{y}{r} + V_{offset}. \tag{3.8}$$

Sensor Fabrication

Aerosol Jet printing technique, also known as maskless mesoscale material deposition (M^3D) [46], is utilized in this work to print the strain gauges on the BCB/PI substrate (cf. Fig. 3.10b) [47–49]. The high-velocity silver-ink jet is capable of creating a compact and high resolution printing with a minimum feature size of 10 μm. The printed structures are cured at a temperature of 150°C.

One of the shortcomings of this printing technique is the mismatched horizontal and vertical printing. This traces back to the accumulation of minor impurities near the nozzle tip during the printing process. Consequently, a mismatch between the resistance values of the vertically and horizontally printed serpentine-shaped structures (R_V and R_H) is observed, which introduces an offset voltage V_{offset} between the two output voltages of the strain gauge bridge circuit (cf. the resistance mismatch of the on-foil RTDs in Sect. 3.1). Selective printing is used to lower the resistance mismatch by following a print-measure-print routine.

Sensor Characterization

After the printing and curing processes are finished, the strain gauges' foil can be released from the rigid carrier wafer. The sensors are then assembled on double-sided foam adhesive that is mounted on a 3D printed plastic material for robotic applications [47–49]. A column with a movable rod is used to incrementally bend the strain gauges. Using a high-resolution camera, side-view images are captured, which are then used to estimate the bending curvature of the strain gauges.

Figure 3.11a shows the measured strain gauge differential output voltage $V_1 - V_2$ with the bending curvature [47–49]. The bridge output voltage is increasing monotonically with increasing the bending curvature (the relationship is ideally linear, cf. Eq. (3.8)). The measured output voltage is about 5mV at a bending radius of 38mm and using 1mA as a biasing current. At no bending, a non-zero output voltage is observed, which traces back to the resistance mismatch between the bridge resistors. As shown in Fig. 3.11b, the memory effect in both the 3D printed plastic material and foam adhesive materials causes the variation of the offset voltage after each bending cycle. This memory effect can be reset if the strain gauge is forced to the no-bending position or left without bending activities for long periods.

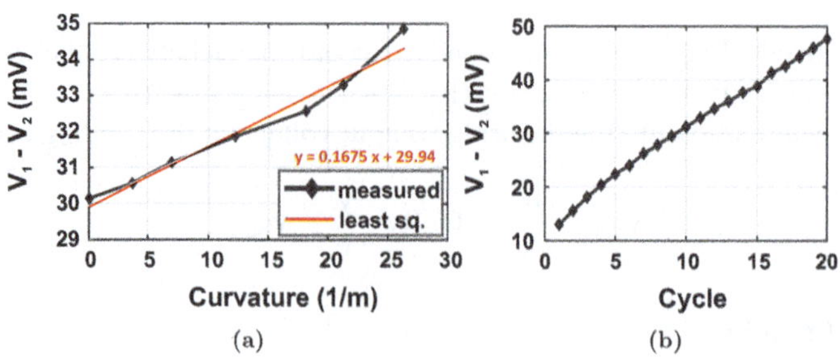

Fig. 3.11 (a) Measurements show a monotonic increase of strain gauge output voltage with increasing the bending curvature [47–49]. (b) Bridge offset voltage increases after each bending cycle, until the memory effect is reset

3.4 On-Foil Spiral Inductors

Flexible sensor systems increasingly require wireless power and data transmission systems, which partly incorporate antennas that are mechanically and electrically tolerant to high levels of bending and stretching. The performance of flexible antennas (e.g. an inkjet-printed 1.2-3.4 GHz multiband antenna [50], 915-MHz antenna fabricated using conductive fibers in PDMS substrate [51] and 5.5-GHz antenna on BCB/PI with embedded transceiver UTC [52, 53]) is steadily approaching those fabricated on conventional rigid substrates.

In this section, off-chip/on-foil spiral inductors are designed, fabricated and characterized that are compatible with the CFP process. Here, the design targets HySiF integration as a part of a flexible NFC sub-system in a multistandard high-performance smart label. However, the approach presented here applies to other HySiF integration scenarios that incorporate spiral inductors. Bendable NFC transceivers [54], energy harvesting systems [55–58], LC wireless sensor readout [59], magnetic resonance imaging (MRI) [60] and proximity sensing [61] are some examples where conventional off-chip planar spiral inductors are needed.

Spiral Inductor Design

Unlike most high-frequency antennas that transmit and receive electromagnetic wave signals, NFC antennas rely mainly on the magnetic field of conductor loops. According to classical electromagnetism, the electric current in a conductor results in a surrounding magnetic field that forms concentric circles of field lines, which decay with increasing the distance from the conductor. The opposite is also true as described by the Maxwell-Faradays' law, as when a time-varying magnetic field induces an electric field. In RFID systems, both scenarios take place as the transmitter coil generates a strong magnetic field that is capable of reaching nearby coil(s) of the passive receiver(s). The transmitter and receiver coils are now inductively coupled, and the mutual magnetic field modulates a voltage difference on a resistor in the receiver circuitry. This voltage is usually rectified to serve as a power supply for the passive NFC tag. It is also used to communicate between the transmitter and receiver in a so-called load-modulation.

NFC systems uses the world-wide ISM-band (industrial, scientific and medical) of 13.56MHz but for short-range secure payments or in cases of multiple tag stacking, the NFC system can be detuned to 14.5 MHz, or even higher frequencies of 17-19MHz [63, 64]. International NFC standards (e.g. ISO/IEC 13157, 14443, 180092, 21481, etc.) define the tag physical characteristics (i.e. antenna size), communication parameters (e.g. modulation and channel coding schemes), the card activation sequences and the digital protocol [65]. In this work, the NFC antenna dimensions, defined by the ISO/IEC 14443, are used from which classes 1 and 3 (Fig. 3.12) are chosen as these classes are most suitable for smart tablets and smart phones.

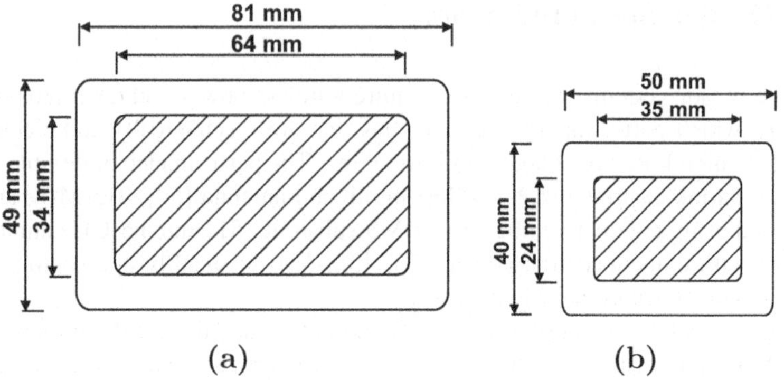

Fig. 3.12 NFC antenna design based on ISO/IEC 14443 for (a) class 1 and (b) class 3 [62]

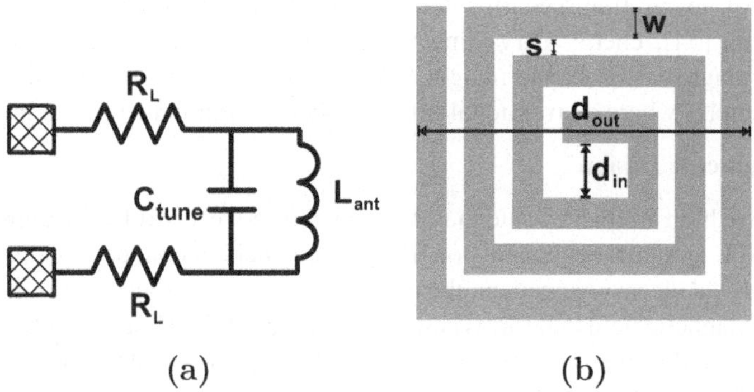

Fig. 3.13 **a** NFC antenna circuit model. **b** Planar spiral inductor design paramters [66]

Depending on the used NFC tag IC and its output impedance, the antenna matching circuit is constructed. In this work, a NFC IC from the company em microelectronics, which is later back-thinned to a thickness of 30μm, is utilized [63]. It is optimized for an operating nominal resonant frequency of 14.5MHz supporting data rates from 106 kbps up to 848 kbps. In addition, it includes an on-chip resonant capacitor of 14pF and its evaluation board includes an off-chip tuning capacitor of 33pF. As shown in Fig. 3.13a, the antenna is modeled using a RLC circuit. If a fixed antenna design is used, the off-chip tuning capacitor C_{tune} adjusts the LC resonance frequency to the nominal operating frequency using the following relation:

$$C_{tune} = \frac{1}{(2 \cdot \pi \cdot f_0)^2 L_{ant}} - C_{chip}, \tag{3.9}$$

where f_0 is the NFC operating frequency, L_{ant} is the antenna inductance and C_{chip} is the on-chip resonant capacitor.

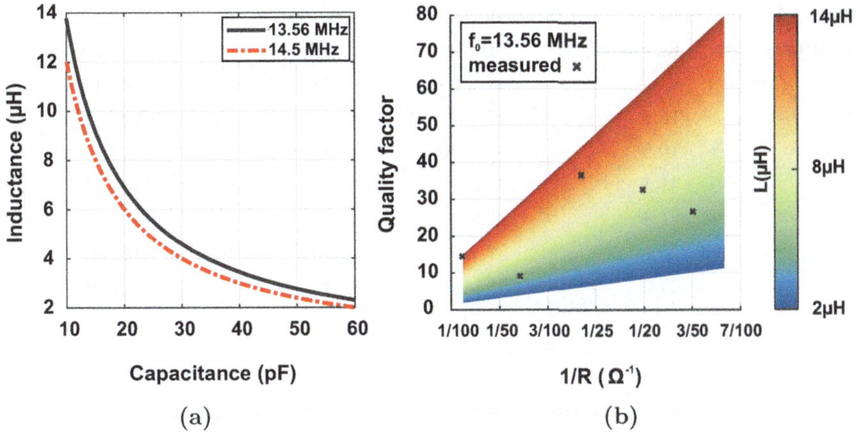

Fig. 3.14 **a** Inductance value plotted against the required total capacitance for two NFC operating frequencies using equation 3.9. **b** Quality factor plane plotted against the loop equivalent conductance for different inductance values using equation 3.10 and some measured on-foil inductors are plotted

In this work, the total capacitance value is limited to a few tens of pF, which originates from the capability of achieving off-chip/on-foil capacitors in this range (cf. Sect. 3.2.2). In order to estimate the inductance of the NFC antenna, equation (3.9) is plotted in Fig.3.14a, where the total capacitance value is swept from 10 to 60 pF for two operating frequencies of 13.56 and 14.5 MHz. Inductance values in the range of 2–14 µH are calculated and these inductance values are used to design planar spiral inductors.

In general, the operating bandwidth (BW) is inversely proportional to the quality factor ($Q = f_0/BW$). Therefore, the Q factor is chosen not too high ($<$ 40 [65]) so it limits the transmission data rate and not too low so the antenna requires more current to be driven by the NFC IC. In general, the Q factor of inductors is limited by the resistive losses in the spiral coil and by the substrate losses [67]. Here, the substrate losses are negligible compared to the resistive loses as the spiral coil is fabricated on BCB polymer, which have low loss tangent of 0.0008 at 1GHz [68]. The Q factor can be calculated using the following relation:

$$Q = \frac{im(Z)}{re(Z)} = \frac{2 \cdot \pi \cdot f_0 \cdot L_{ant}}{R_L}, \tag{3.10}$$

where Z is the coil impedance, R_L is the total series resistance of the coil conductor in addition to any additional circuit interconnections. A widely used and best-practiced R_L range is from 20 to 80 Ω [65]. Figure 3.14b plots the inverse resistance in the previously mentioned range against the Q factor for different inductance values that result from Fig. 3.14a.

Table 3.2 Spiral inductors measured performance summary [69]

Inductance (μH)	Resistance (Ω)	Size (mm2)	Turns	s, w (μm)
3.6	9.2	41 × 55	5	600, 300
5.2	16.6	41 × 71	5	400, 400
3.3	10.5	35 × 40	5	400, 200
4.4	41.3	25 × 35	5	100, 100
13.8	81.3	25 × 35	10	100, 100
7.7	20.1	35 × 40	10	400, 200
11.6	27	35 × 40	15	400, 200

Figure 3.13b shows the different geometric parameters such that the number of turns n, turn width w, turn spacing s, outer diameter d_{out}, inner diameter d_{in}, average diameter $d_{avg} = 0.5(d_{out} + d_{in})$, fill ratio $\rho = (d_{out} - d_{in})/(d_{out} + d_{in})$ are defined. The planar spiral inductance values are calculated using the modified Wheeler empirical formula [66] as follows:

$$L = K_1 \mu_0 \frac{n^2 \cdot d_{avg}}{1 + K_2 \cdot \rho}, \tag{3.11}$$

where the coefficients K_1 and K_2 equals to 2.34 and 2.75 for spiral coils with square layout and μ_0 is the electrical permeability, which is defined as $4\pi \cdot 10^{-7}$ H/m. The metal sheet resistance together with the desired resistance and inductance determine the number of turns n and turn width w and the capacitive coupling between each turn limits the turn spacing s. Different spiral coils are designed and their parameters and characterization results are listed in Table 3.2.

Spiral Inductor Fabrication

Here, spiral inductors are fabricated on BCB/PI substrate that is similar to the CFP package stack [69]. Figure 3.15 shows the 150-mm wafer with the fabricated spiral inductors, the structure of which follows the ISO/IEC 14443 standard for NFC antennas [62]. Following the CFP process, a BCB layer is spin-coated, which has a thickness of 5 μm and cured at 250°. An underpass is then fabricated using AlSiCu sputtering having a thickness of 1 μm. Note, that for lower parasitic resistance, a double-layer spiral inductor can be fabricated where the first layer is fabricated in the same processing step as the underpass.

After the fist metalization step, another 1-3-μm thick layer of BCB is spin-coated and cured. Using standard lithography, array of multiple CFP-standard vias (size = 30 μm× 30 μm) is opened to connect the underpass with the inductor main loop of the top metal. Another AlSiCu metalization step is done for achieving the top

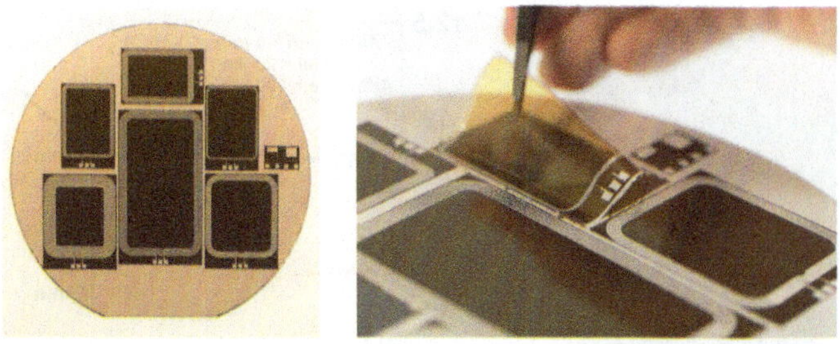

(a) (b)

Fig. 3.15 (a) Photograph of a test wafer for the fabricated on-foil spiral inductors using BCB/PI substrate [69]. (b) Photograph demonstrating the mechanical detachment of a spiral inductor from the carrier wafer

metal and main inductor loop with a thickness of 2 μm. The inductors are then released from the carrier wafer as illustrated in Fig. 3.15b. Note, that as the metal trace density increases, the stress induced on the flexible substrate increases, which leads to warpage and difficulties during the foil release [69].

Spiral Inductor Characterization

Table 3.2 summarizes the electrical characterization results of the on-foil planar spiral inductors [69] and Q factor is previously plotted in Fig. 3.14b. The electrical characterization is performed by connecting the inductor under test to LCR meter (MFIA Impedance Analyzer) using four-point connection.

Figure 3.16b shows the measured inductance of a spiral inductor (25×25 mm^2, 10 turns, $w = s = 100\mu m$) [69]. The tensile uniaxial tensile strain acting on the inductor increases with increasing the bending radii. As a result, the spiral loop average diameter d_{avg} increases, which increases the inductance values (cf. Eq. (3.11) and [70]). The average diameter, calculated as $d_{avg} = 0.5(d_{out} + d_{in})$, increases since the inductor traces expand upon tensile uniaxial strain and with small Poisson ratio for metals, both d_{out} and d_{in} increases.

3.5 Thin-Film Organic Electronics

Previous work has demonstrated in-depth device modeling [71] and the realization of OTFT-based complex circuits, such as shift registers [72], ADCs [73], DACs [74] and digital processor [75]. However, such circuits usually require high voltages to operate in the range of 10–20 V.

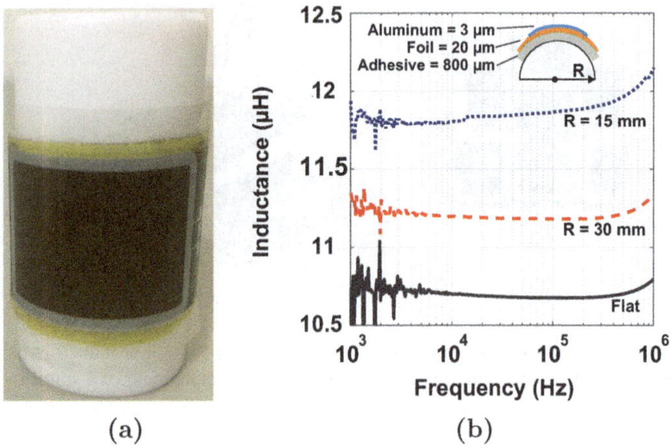

Fig. 3.16 (a) Photograph of the spiral inductor during bending [69]. (b) Characterization of the spiral inductor for different bending radii

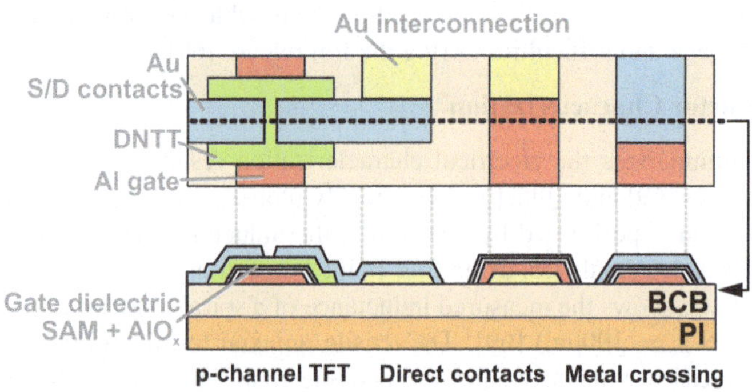

Fig. 3.17 Top and cross-sectional views of the OTFTs on the flexible BCB/PI substrate [76]

Fabrication

Using high-resolution silicon stencil masks, the inverted staggered OTFTs are fabricated on the flexible BCB/PI substrate [71, 76]. As shown in Fig. 3.17, the organic transistors use the p-channel organic semiconductor dinaphtho[2,3-b:2′,3′-f]thieno[3,2-b]thiophene (DNTT) and have channel lengths as short as 5 µm and gate-to-contact overlaps of 20µm. The gate electrodes and circuit interconnects of the first layer are fabricated first using a 30-nm-thick of Al by thermal evaporation in vacuum. The gate dielectric is a stack of 3.6-nm thick AlO_x layer and 1.7-nm-thick solution-processed tetradecylphophonic acid self-assembled monolayer (SAM). The capacitance of the hybrid gate dielectric is about 700 nF/cm^2 [77] that allows for low-voltage operation (3.3–2 V). A 25-nm-thick layer of the small-molecule organic semiconductor DNTT is then deposited. In order to fabricate the source/drain contacts and circuit interconnects of the second layer, a 25-nm-thick Au is evaporated.

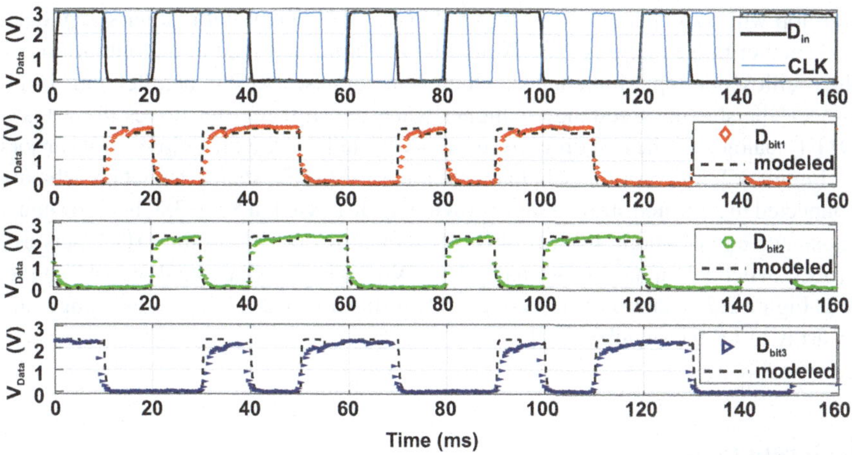

(a)

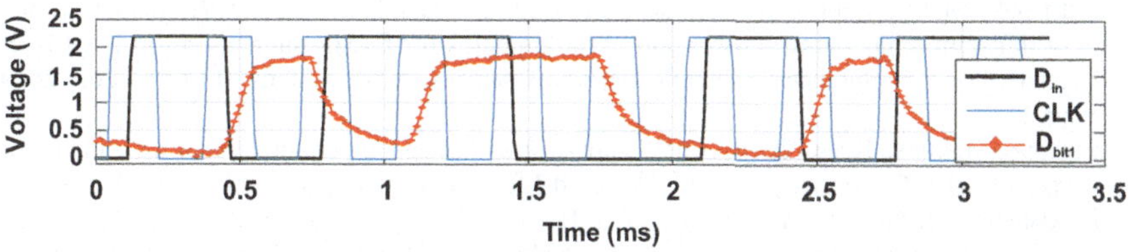

(b)

Fig. 3.21 **a** Measurement and modeling of a 3-stage shift register operated using a supply voltage of 3V, bias voltage of -1V and clock frequency of 100 Hz. **b** Measurements of 1-stage shift register operated using supply voltage of 2.2 V, bias voltage of −1 V and clock frequency of 3 kHz

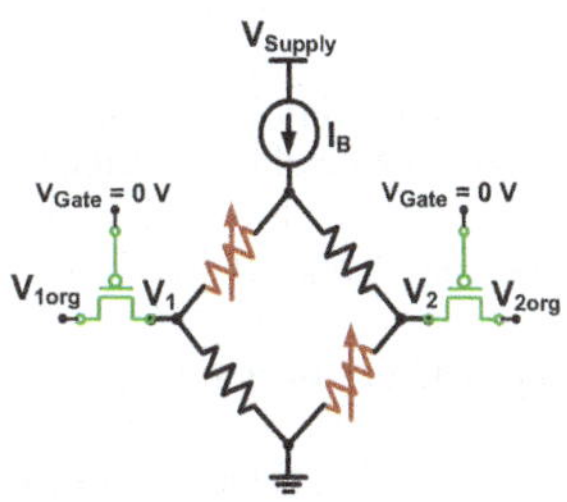

(a)

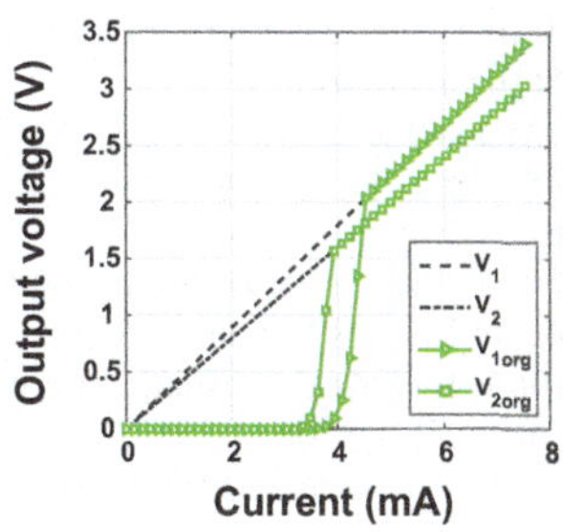

(b)

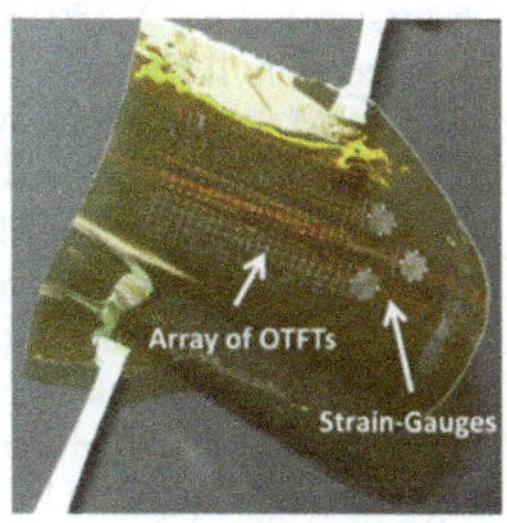

(c)

Fig. 3.22 **a** Schematic of one strain gauge biased using current source and connect to a pair of OTFT analog switches. **b** Static measurements showing the performance of the OTFT analog switches in the off/on states. **c** OTFTs and printed strain gauges fabricated on the surface of the BCB/PI substrate [82] flip-flop will also operate with a bias voltage of 0 V, albeit with larger signal delay and smaller noise margins [76]. Note, that the minimum load at the output node is the oscilloscope input impedance, which is in our case 1MΩ in parallel with 16 pF. In fact, the shifting frequency is higher when the shift register drives the gates of OTFT analog switches, in order to successively address the off-chip/on-foil sensors.

Figure 3.22a shows the schematic of a pair of OTFT analog switches, which is connected to a printed strain gauge (previously discussed in Sect. 3.3, here the gauge resistance equals about 400 Ω) [49]. As shown in Fig. 3.22b, the OTFT acts as a proper switch when its gate-source voltage V_{GS} is higher than the threshold voltage V_{th}. Figure 3.22c shows the released BCB/PI substrate including 3 strain gauges and an array of OTFT switches [82].

References

1. Elsobky M, Ottaviani A, Alomari M, Yu Z, Deuble T, Burghartz J-N (2019) Characterization of thin-film temperature sensors and ultra-thin chips for HySiF integration. In: 2019 IEEE international conference on flexible and printable sensors and systems (FLEPS), July 2019, pp 1–3
2. Pan S, Makinwa KAA (2018) A 0.25 mm^2-resistor-based temperature sensor with an inaccuracy of 0.12 °C (3σ) from −55 to 125 °C. IEEE J Solid-State Circ 53(12):3347–3355
3. Alshahed M, Yu Z, Rempp H, Richter H, Harendt C, Burghartz J-N (2014) Thermal characterization and modeling of ultra-thin silicon chips. In: 2014 44th European solid state device research conference (ESSDERC), Sep 2014, pp 397–400
4. Yao Y, Lu G, Boroyevich D, Ngo KDT (2015) Survey of high-temperature polymeric encapsulants for power electronics packaging. IEEE Trans Components Pack Manuf Technol 5(2):168–181
5. Ahmed M, Chitteboyina MM, Butler DP, Celik-Butler Z (2012) Temperature sensor in a flexible substrate. IEEE Sens J 12(5):864–869
6. Hsiao F, Liao Y (2018) Printed micro-sensors for simultaneous temperature and humidity detection. IEEE Sens J 18(16):6788–6793
7. Soni M, Bhattacharjee M, Manjakkal L, Dahiya R (2019) Printed temperature sensor based on graphene oxide/PEDOT:PSS. In: 2019 IEEE international conference on flexible and printable sensors and systems (FLEPS), July 2019, pp 1–3
8. Widlund T, Yang S, Hsu Y-Y, Lu N (2014) Stretchability and compliance of freestanding serpentine-shaped ribbons. Int J Solids Structures 51:08
9. Parker RL, Krinsky A (1963) Electrical resistance-strain characteristics of thin evaporated metal films. J App Phys 34(9):2700–2708
10. Baetens T, Pallecchi E, Thomy V, Arscott S (2018) Cracking effects in squashable and stretchable thin metal films on PDMS for flexible microsystems and electronics. Sci Rep 8:12
11. Modafe A, Ghalichechian N, Powers M, Khbeis M, Ghodssi R (2005) Embedded benzocyclobutene in silicon: an integrated fabrication process for electrical and thermal isolation in mems. Microelectron Eng 82(2):154–167. Available http://www.sciencedirect.com/science/article/pii/S0167931705003175
12. Malik IJ, Pirooz S, Shive LW, Davenport AJ, Vitus CM (1993) Surface roughness of silicon wafers on different lateral length scales. J Electrochem Soc 140(5):75–77
13. Bosch-Sensortec Homepage (2019) Available https://www.bosch-sensortec.com/bst/products/environmental/overview_environmental
14. Sensirion Homepage (2019) Available https://www.sensirion.com/environmental-sensors/

15. Ghosh S, Ghosh R, Guha P, Bhattacharyya T (2015) Humidity sensor based on high proton conductivity of graphene oxide. IEEE Trans Nanotechnol 14:931–937
16. Leng X, Li W, Luo D, Wang F (2017) Differential structure with graphene oxide for both humidity and temperature sensing. IEEE Sens J 17(14):4357–4364
17. Szendrei K, Ganter P, Sánchez-Sobrado O, Eger R, Kuhn A, Lotsch B (2015) Touchless optical finger motion tracking based on 2D nanosheets with giant moisture responsiveness. Adv Mater 27(09) (Deerfield Beach, FL)
18. Morais RM, Klem MdS, Nogueira GL, Gomes TC, Alves N (2018) Low cost humidity sensor based on PANI/PEDOT:PSS printed on paper. IEEE Sens J 18(7):2647–2651
19. Mogera U, Sagade AA, George JS, Kulkarni UG (2014) Ultrafast response humidity sensor using supramolecular nanofibre and its application in monitoring breath humidity and flow. Sci Rep 4(14): 4103
20. Chen Z, Lu C (2005) Humidity sensors: a review of materials and mechanisms. Sens Lett 3:12
21. Elsobky M, Albrecht B, Richter H, Burghartz JN, Ganter P, Szendrei K, Lotsch BV (2017) Ultra-thin relative humidity sensors for hybrid system-in-foil applications. In: 2017 IEEE sensors, pp 1–3
22. Randles JEB (1947) Kinetics of rapid electrode reactions. In: Discussions of the Faraday society, pp 11–19
23. Gore WL Gore & Associates GmbH Homepage (2019) Available https://www.gore.de/products/gore-protective-vents-speziell-fuer-leuchtengehaeuse
24. Tan Z, Daamen R, Humbert A, Ponomarev YV, Chae Y, Pertijs MAP (2013) A 1.2-V 8.3-nJ CMOS humidity sensor for RFID applications. IEEE J Solid-State Circ 48(10):2469–2477
25. Rivadeneyra A, Fernández-Salmerón J, Agudo-Acemel M, López-Villanueva JA, Capitan-Vallvey LF, Palma AJ (2016) Printed electrodes structures as capacitive humidity sensors: a comparison. Sens Actuat A: Phys 244:56–65. Available http://www.sciencedirect.com/science/article/pii/S0924424716301224
26. Boudaden J, Steinmaßl M, Endres HE, Drost A, Eisele I, Kutter C, Müller-Buschbaum P (2018) Polyimide-based capacitive humidity sensor. Sensors (Basel, Switzerland) 18
27. Boltshauser T, Leme CA, Baltes H (1993) High sensitivity CMOS humidity sensors with on-chip absolute capacitance measurement system. Sens Actuat B: Chem 15(1):75–80. Available http://www.sciencedirect.com/science/article/pii/092540059385030E
28. Ha H, Sylvester D, Blaauw D, Sim J (2014) 12.6 A 160 nW 63.9 fJ/conversion-step capacitance-to-digital converter for ultra-low-power wireless sensor nodes. In: 2014 IEEE international solid-state circuits conference digest of technical papers (ISSCC), pp 220–221
29. Ghanbari MM, Tsai JM, Nirmalathas A, Muller R, Gambini S (2017) An energy-efficient miniaturized intracranial pressure monitoring system. IEEE J Solid-State Circ 52(3):720–734
30. Xie M, Wang L, Dong L, Deng W, Huang Q (2019) Low cost paper-based LC wireless humidity sensors and distance-insensitive readout system. IEEE Sens J 19(12):4717–4725
31. Fujifilm Homepage (2019) Available https://www.fujifilm.com/products/semiconductor_materials/
32. Hassan M, Schomburg C, Harendt C, Penteker E, Burghartz J-N (2013) Assembly and embedding of ultra-thin chips in polymers. In: 2013 Eurpoean microelectronics packaging conference (EMPC), Sep 2013, pp 1–6
33. Laconte J, Wilmart V, Raskin J, Flandre D (2003) Capacitive humidity sensor using a polyimide sensing film. In: Symposium on design, test, integration and packaging of MEMS/MOEMS 2003, May 2003, pp 223–228
34. Schubert PJ, Nevin JH (1985) A polyimide-based capacitive humidity sensor. IEEE Trans Electron Devices 32(7):1220–1223
35. Igreja R, Dias C (2004) Analytical evaluation of the interdigital electrodes capacitance for a multi-layered structure. Sens Actuators Phys 291–301
36. Denton DD, Day DR, Priore DF, Senturia SD, Anolick ES, Scheider D (1985) Moisture diffusion in poltimide films in integrated circuits. J Electron Mater 14(2):119–136. Available https://doi.org/10.1007/BF02656671

37. Zhu Z, Yang G, Li R, Pan T (2017) Photopatternable PEDOT:PSS/PEGhybrid thin film with moisture stability and sensitivity. Microsyst Nanoeng 3:17004
38. Bi H, Yin K, Xie X, Ji J, Wan S, Sun L, Terrones M, Dresselhaus M (2013) Ultrahigh humidity sensitivity of graphene oxide. Sci Rep 3:2714
39. Borini S, White R, Wei D, Astley M, Haque S, Spigone E, Harris N, Kivioja J, Ryhänen T (2013) Ultrafast graphene oxide humidity sensors. ACS Nano 7:11
40. Guo H, Lan C, Zhou Z, Sun P, Wei D, Li C (2017) Transparent, flexible, and stretchable WS_2 based humidity sensors for electronic skin. Nanoscale 9:05
41. Jingwen Q, Peng Z, Shen Z, Zhao Z, Zhang G, Fu X (2016) Positive impedance humidity sensors via single-component materials. Sci Rep 6:25574
42. Phan D-T, Park I, Park A-R, Park C-M, Jeon K-J (2017) Black p/graphene hybrid: a fast response humidity sensor with good reversibility and stability. Sci Rep 7:12
43. Khan MU, Hassan G, Bae J (2019) Bio-compatible organic humidity sensor based on natural inner egg shell membrane with multilayer crosslinked fiber structure. Sci Rep 9:04
44. Cai D, Tong T, Zhang Z, Pan J, Zhang L, Tong L (2019) Functional film coated optical micro/nanofibers for high-performance gas sensing. IEEE Sens J 19(20):9229–9234
45. Innovative Sensor Technology IST AG Homepage (2019). Available https://www.ist-ag.com/
46. Hedges M, Renn M, Kardos M (2005) Mesoscale deposition technology for electronics applications. 2005(11): 53–57
47. Elsobky M (2015) Design of a flexible smart skin for robotic application. Master's thesis, University of Ulm
48. Elsobky M, Mahsereci Y, Keck J, Richter H, Burghartz J (2017) Design of a CMOS readout circuit on ultra-thin flexible silicon chip for printed strain gauges. Adv Radio Sci 15:123–130
49. Elsobky M, Mahsereci Y, Yu Z, Richter H, Burghartz JN, Keck J, Klauk H, Zschieschang U (2018) Ultra-thin smart electronic skin based on hybrid system-in-foil concept combining three flexible electronics technologies. Electron Lett 54(6):338–340
50. Ahmed S, Tahir FA, Shamim A, Cheema HM (2015) A compact Kapton-based inkjet-printed multiband antenna for flexible wireless devices. IEEE Antennas Wirel Propag Lett 14:1802–1805
51. Kiourti A, Volakis JL (2014) Stretchable and flexible e-fiber wire antennas embedded in polymer. IEEE Antennas Wirel Propag Lett 13:1381–1384
52. Meister T, Ellinger F, Bartha JW, Berroth M, Burghartz J, Claus M et al (2017) Program FFlexCom—high frequency flexible bendable electronics for wireless communication systems. In: 2017 IEEE international conference on microwaves, antennas, communications and electronic systems (COMCAS), pp 1–6
53. Alavi G (2019) Hybrid system-in-foil integration and interconnection technology based on adaptive layout technique. Ph.D. dissertation, University of Stuttgart
54. Myny K, Lai Y, Papadopoulos N, De Roose F, Ameys M, Willegems M, Smout S, Steudel S, Dehaene W, Genoe J (2017) 15.2 A flexible ISO14443-a compliant 7.5 mW 128 b metal-oxide NFC barcode tag with direct clock division circuit from 13.56 MHz carrier. In: 2017 IEEE international solid-state circuits conference (ISSCC), pp 258–259
55. Miura N, Mizoguchi D, Sakurai T, Kuroda T (2005) Analysis and design of inductive coupling and transceiver circuit for inductive inter-chip wireless superconnect. IEEE J Solid-State Circ 40(4):829–837
56. Hu Y, Rieutort-Louis W, Huang L, Sanz-Robinson J, Wagner S, Sturm JC, Verma N (2012) Flexible solar-energy harvesting system on plastic with thin-film LC oscillators operating above ft for inductively-coupled power delivery. In: Proceedings of the IEEE 2012 custom integrated circuits conference, Sep 2012, pp 1–4
57. Kim J, Salvatore GA, Araki H, Chiarelli A, Xie Z, Banks A, Sheng X, Liu Y, Lee J, Jang KI, Heo S, Cho K, Luo H, Zimmerman B, Yan L, Feng X, Xu S, Fabiani M, Gratton G, Rogers J (2016) Battery-free, stretchable optoelectronic systems for wireless optical characterization of the skin. Sci Adv 2:e1 600 418–e1 600 418
58. Lo C, Cheng H, Liao P, Chen Y, Chen P (2018) An 82.1%-power-efficiency single-inductor triple-source quad-mode energy harvesting interface with automatic source selection and

reversely polarized energy recycling. In: 2018 IEEE Asian solid-state circuits conference (A-SSCC), Nov 2018, pp 165–168
59. Huang Q, Dong L, Wang L (2016) LC passive wireless sensors toward a wireless sensing platform: status, prospects, and challenges. J Microelectromech Syst 25(5):822–841
60. Handwerker J, Eder M, Tibiletti M, Rasche V, Scheffler K, Becker J, Ortmanns M, Anders J (2016) An array of fully-integrated quadrature TX/RX NMR field probes for MRI trajectory mapping. In: ESSCIRC conference 2016: 42nd European solid-state circuits conference, Sep 2016, pp 217–220
61. Kamotesov S, Lombard P, Vollaire C, Semet V, Cabrera M, Dahmani R, Veille A, Moguedet M (2016) Modelization and characterization of 2D and 3D mid inductors for multidirectional inductive proximity sensing. In: 2016 12th International congress molded interconnect devices (MID), Sep 2016, pp 1–6
62. •, [ISO/IEC 14443-4:2018] Cards and security devices for personal identification— Contactless proximity objects. International Organization for Standardization, ISO, Technical Report, 2018
63. Em Microelectronic Homepage (2019). Available https://www.emmicroelectronic.com/
64. Paret D (2016) Design constraints for NFC devices. Wiley
65. Training MK (2015) NFC reader design: antenna design considerations. NXP semiconductors, Technical report. http://miniradiosolutions.com/wp-content/uploads/2015/09/NFC-Reader-Design-II-Antenna-design-considerations-Public.pdf
66. Mohan SS, del Mar Hershenson M, Boyd SP, Lee TH (1999) Simple accurate expressions for planar spiral inductances. IEEE J Solid-State Circ 34(10):1419–1424
67. Burghartz JN, Edelstein DC, Soyuer M, Ainspan HA, Jenkins KA (1998) RF circuit design aspects of spiral inductors on silicon. IEEE J Solid-State Circ 33(12):2028–2034
68. Di Massa G, Costanzo S, Borgia A, Venneri F, Venneri I (2010) Innovative dielectric materials at millimeter-frequencies. In: 2010 Conference proceedings ICECom, 20th international conference on applied electromagnetics and communications, Sep 2010, pp 1–4
69. Elsobky M, Alavi G, Albrecht B, Deuble T, Harendt C, Richter H, Yu Z, Burghartz J (2018) Ultra-thin sensor systems integrating silicon chips with on-foil passive and active components. In: Proceedings, vol 2, p 748
70. Fassler A, Majidi C (2013) Soft-matter capacitors and inductors for hyperelastic strain sensing and stretchable electronics. Smart Mater Structures 22(5):055023. Available https://doi.org/10.1088
71. Zaki T (2015) Short-channel organic thin-film transistors: fabrication, characterization, modeling and circuit demonstration. Ph.D. dissertation, University of Stuttgart, Springer
72. Uno M, Isahaya N, Cha B-S, Omori M, Yamamura A, Matsui H, Kudo M, Tanaka Y, Kanaoka Y, Ito M, Takeya J (2017) High-yield, highly uniform solution-processed organic transistors integrated into flexible organic circuits. Adv Electron Mater 3:1600410
73. Marien H, Steyaert MSJ, van Veenendaal E, Heremans P (2011) A fully integrated $\Delta\Sigma$ ADC in organic thin-film transistor technology on flexible plastic foil. IEEE J Solid-State Circ 46(1):276–284
74. Zaki T, Ante F, Zschieschang U, Butschke J, Letzkus F, Richter H, Klauk H, Burghartz JN (2012) A 3.3 V 6-Bit 100 kS/s current-steering digital-to-analog converter using organic p-type thin-film transistors on glass. IEEE J Solid-State Circ 47(1):292–300
75. Myny K, van Veenendaal E, Gelinck GH, Genoe J, Dehaene W, Heremans P (2012) An 8-bit, 40-instructions-per-second organic microprocessor on plastic foil. IEEE J Solid-State Circ 47(1):284–291
76. Elsobky M, Elattar M, Alavi G, Letzkus F, Richter H, Zschieschang U, Strecker M, Klauk H, Burghartz J (2017) A digital library for a flexible low-voltage organic thin-film transistor technology. Organ Electron 50:491–498
77. Kraft U, Anthony JE, Ripaud E, Loth MA, Weber E, Klauk H (2015) Low-voltage organic transistors based on tetraceno[2,3-b]thiophene: contact resistance and air stability. Chem Mater 27(3):998–1004

78. Elsobky M (2013) Measurements and characterization of low-voltage organic-based transistors and digital logic gates. Bachelor thesis, University of Stuttgart
79. Hauser JR (1993) Noise margin criteria for digital logic circuits. IEEE Trans Educ 36(4):363–368
80. Huang T, Fukuda K, Lo C, Yeh Y, Sekitani T, Someya T, Cheng K (2011) Pseudo-CMOS: a design style for low-cost and robust flexible electronics. IEEE Trans Electron Devices 58(1):141–150
81. Elattar M (2014) Design of an organic-based driver circuit for smart skin application. Bachelor thesis, University of Stuttgart
82. Hassan M, Keck J, Klauk H, Kostelnik J, Mahsereci Y, Sailer S, Schreivogel A, Zaki T, Burghartz JN (2015) Combining organic and printed electronics in hybrid system in foil (HySiF) based smart skin for robotic applications. In: 2015 European microelectronics packaging conference (EMPC), Sep 2015, pp 1–6

Chapter 4
Ultra-Thin Chips: Sensors Readout and Microcontrollers

Due to the strengthened requirements on material purity and uniformity, single-crystalline silicon has a limited surface area coverage, which is governed by the maximum available wafer size. Ultra-Thin Chips (UTCs) are considered the representative of the mature silicon CMOS industry in the field of flexible electronics, thus unlocking various applications where embedded intelligence and data processing are required. In this chapter, the potential of UTCs as a promising candidate and building block for high-performance smart flexible electronic applications is investigated. Four case studies are designed to evaluate the impact of the chip back-thinning on the performance of different UTCs and propose circuit design methodologies, which address key HySiF challenges. The chapter starts with the implementation of a smart humidity sensor system, which uses readout and microcontroller UTCs and off-chip/on-foil humidity sensor. Here, the readout circuit is discussed within the HySiF concept and the impact of the readout chip thinning is investigated. Next, a smart resistive readout system, which cancels process-induced errors of printed strain gauges, is discussed. In addition, an Application-Specific Integrated Circuit (ASIC) prototype is implemented in a 0.18 μm CMOS technology. Multiple readout channels and an integrated temperature sensor are implemented for interfacing with off-chip/on-foil sensors. Finally, the focus is directed towards the electrical and thermal characterization of ultra-low-power microcontroller UTCs.

4.1 Humidity Sensor System

The research in capacitance-to-digital converters (CDCs) is mainly driven by the ever-increasing need for precision MEMS-based capacitive sensors. In this section, a CDC ASIC is utilized as a readout interface circuit for the previously discussed ultra-thin capacitive relative humidity sensor (cf. Chap. 3, Sect. 3.2).

M. Elsobky, *Ultra-Thin Sensors and Data Conversion Techniques for Hybrid System-in-Foil*, Recognizing Outstanding Ph.D. Research,
https://doi.org/10.1007/978-3-030-97726-9_4

System Design

CMOS-integrated CDCs can be divided into two categories namely, direct CDCs that incorporate the sense capacitor within an ADC or oscillator implementation and indirect CDCs in which a capacitance-to-voltage or capacitance-to-current converters (CVC, CCC) is followed by a voltage-mode or current-mode ADC. Various ADC architectures are adjusted for a direct CDC implementation. For example, the sense capacitor is connected in parallel to the CDAC in a SAR-based (Successive Approximation Register) ADC implementing a power-efficient CDC [1]. However, the parasitic capacitance and the loss of dynamic range (DR) due to the inevitable quantization of the sensor base capacitance value are the main challenges. Another precision CDC uses a similar SAR-based implementation followed by a fine conversion using an incremental $\Sigma\Delta$ ADC achieving ENOB (Effective Number of Bits) of 15.4b [2]. However, the CDAC matching requirements and the need for complex calibration are the main challenges. Finally, the sensor capacitor(s) is (are) incorporated in the first SC integrator of $\Sigma\Delta$ ADCs achieving precision CDC operation [3, 4]. This technique overcomes the shortcomings of the previous direct CDC designs, but since the $\Sigma\Delta$ modulator coefficients are defined by the sense capacitor, strengthened matching requirements are imposed. Other direct CDC designs are reported, including period modulation [5] and iterative delay-chain discharging [6], which use time-to-digital conversion techniques.

Indirect CDCs typically use virtual ground node(s) of an Operational Transconductance Amplifier (OTA) to bypass the sensor parasitic capacitance [1]. In this case, the strength of the half or full bridge drivers increase proportional to the parasitic capacitance. Additionally, in voltage-mode CDCs, an SC Programmable Gain Amplifier (PGA) is typically used, followed by a SAR ADC [7]. In current-mode CDCs, a set of current mirrors followed by an integrating dual-slope ADC has been reported [8]. Indirect CDCs are more flexible in terms of DR selectivity, which is needed when interfacing with off-chip sensors.

In this work, indirect voltage-mode CDCs that use SC CVC followed by SAR ADC are utilized [9, 10]. A CDC ASIC that implements charge balancing using an active SC integrator is connected to the off-chip/on-foil ultra-thin RH sensor. Figure 4.1 shows the design of the humidity sensor readout system. Off-chip sense C_s and reference C_r capacitors are connected to the CDC ASIC that was previously designed and fabricated in a 0.5-μm CMOS technology [11]. The CDC ASIC integrates two SC PGAs, one 8-bit SAR ADC and Serial Peripheral Interface (SPI) that communicates with microcontroller unit (MCU). It operates using a supply voltage and clock signal of 5 V and 1 MHz and consumes about 3 mA for a measurement time of 8 μs. The CDC ASIC integrates a SC PGA where the sampling feedback capacitor C_f is implemented on-chip by means of a 6-bit CDAC with a total capacitance value of 5.9 pF. An optional off-chip $C_{f,ex}$ connection is possible in case the on-chip capacitance is not large enough, which can be implemented here by scaling the off-chip reference capacitor in order to set the final CVC gain.

In this basic structure of the CVC, shown in Fig. 4.1, the SC action is controlled using two non-overlapping clock phases Φ_1 and Φ_2. During Φ_1, the feedback

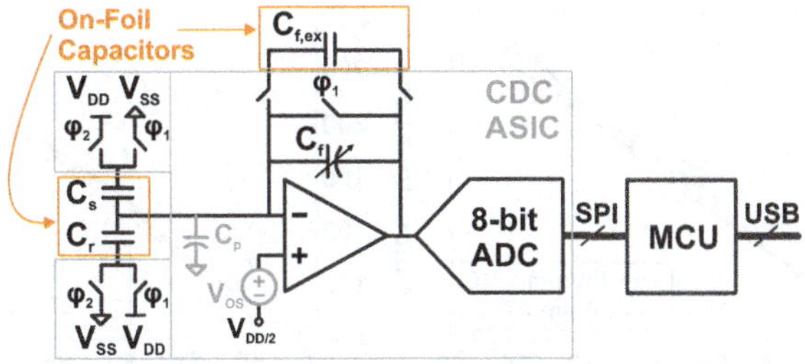

Fig. 4.1 Humidity sensor readout system including a CDC ASIC, MCU and off-chip sense C_s and reference C_r capacitors with an optional off-chip sampling capacitor $C_{f,ex}$. The SC CVC, which includes a half bridge capacitive driver, implements charge balancing between the sense and reference capacitors as the common node is connected to the OTA virtual ground node

capacitor C_f is discharged and the sense and reference capacitors are precharged using bipolar reference voltages $\pm V_{ref}$ or unipolar reference voltage V_{ref} and the ground voltage. The analog supply and ground voltages V_{DD} and V_{SS} are used as the reference voltages. During Φ_2, the charge imbalance between C_s and C_r, which equals $(C_s - C_r)V_{DD}$, flows to the feedback capacitance C_f as the OTA keeps the virtual ground potential constant. The CVC output voltage V_{out} during Φ_2 is given by the following relation:

$$\begin{aligned} V_{out} &= \\ &= \frac{C_s - C_r}{C_f + (C_s + C_r + C_p)/A_0} V_{DD} +, \\ &\frac{(C_s + C_r + C_f + C_p) - A_0(C_s + C_r + C_p)/(A_0 + 1)}{C_f + (C_s + C_r + C_f + C_p)/A_0} V_{OS} + \frac{V_{DD}}{2}, \end{aligned} \quad (4.1)$$

where C_p is the interconnection parasitic capacitance, A_0 is the OTA open loop gain and V_{OS} is the input referred offset of the OTA [12]. The error in the output voltage due to the finite dc gain ϵ_A and the offset voltage scaling factor ϵ_{OS} are used to rewrite Eq. (4.1) in simpler form:

$$V_{out} = \frac{1}{1 + \epsilon_A} \frac{C_s - C_r}{C_f} V_{DD} + \epsilon_{OS} V_{OS} + \frac{V_{DD}}{2}. \quad (4.2)$$

Assuming $A_0 \to \infty$, the relation is simplified and is given by:

$$V_{out} = \frac{C_s - C_r}{C_f} V_{DD} + V_{OS} + \frac{V_{DD}}{2}. \quad (4.3)$$

Here, some disadvantages can be observed. From Eqs. (4.1) and (4.2), the offset voltage appears at the output voltage and ideally it will not be amplified (Assuming Atextsubscript0$\to\infty$, $\epsilon_{OS}\to 1$). In addition, the output and inverting terminals of

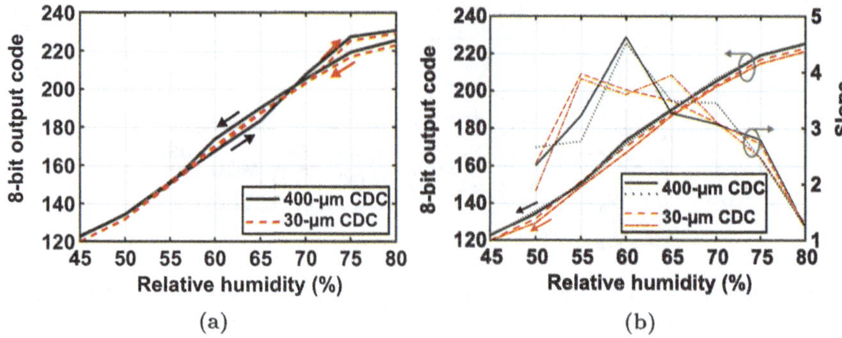

Fig. 4.2 Static characterization of the RH sensor readout system. **a** Measured CDC 8-bit output code plotted versus relative humidity during adsorption and desorption for 400- and 30-μm thick CDC ASICs. **b** Measured CDC 8-bit output code and slope plotted vs. relative humidity during desorption for two 400- and two 30-μm thick CDC ASICs

the OTA, that is incorporated in the basic CVC, are short-circuited for resetting the feedback capacitor. Consequently, the output voltage switches back and forth between the common-mode voltage and the desired output voltage in every clock cycle. Therefore, an OTA with relatively high slew rate (SR) should be implemented. Finally, the circuit accuracy is directly related to the OTA open loop gain A_0 and the error in the output voltage ϵ_A is given by:

$$\epsilon_A = \frac{C_s + C_r + C_p}{A_0 C_f}. \tag{4.4}$$

An enhanced version of this basic SC PGA that is less sensitive to the open loop dc gain A_0 and other parasitic effects is discussed later in Sect. 4.3.2.

Measurement Results

For the humidity sensor readout characterization, the 3-μm thick off-chip/on-foil sense and reference capacitors are detached from the carrier wafer. They are assembled using double-sided foam adhesive on a PCB, where pin headers are soldered to copper contact pads. The electrical connections from the sensor to the PCB are made using silver glue (EPO-TEK H20E). The CDC readout ASICs are back-thinned to conventional and ultra-thin thicknesses of 400 and 30 μm in order to test the ASIC performance after the back-thinning process. Next, the sense and reference capacitors are connected to the CDC readout ASICs, which are assembled in conventional ceramic packages. Finally, the sensor system is placed in a controlled climate chamber (Vötsch VCL 0010) and the RH is varied from 45 to 80% with 5% steps at a constant temperature of 30 °C.

For this characterization, the base capacitance values of the sense C_s and reference capacitors C_r equals 24.3 and 27.6 pF when measured using 4-point connection to an LCR meter at RH of about 55% and temperature of about 21 °C. The on-chip feedback capacitance is not large enough for the wide sensitivity range of the sensor, therefore a 47-pF off-chip ceramic capacitor is connected to the CDC ASICs.

Figure 4.2a shows the output code of two CDC ASICs with thicknesses of 400 and 30 µm during humidity adsorption and desorption (i.e. RH is swept between 40 and 80%). Comparable performance is achieved for both CDC ASICs. Compression is observed at the higher humidity values, which accounts for the inability of the climate-controlled chamber to maintain high levels of humidity. Hysteresis in the CDC response traces back to the turbulent airflow resulting from the always-on fan of the climate-controlled chamber.

Figure 4.2b plots the CDC output code on the left axis for 4 CDC ASICs during desorption (i.e. RH decreasing from 80 to 45%). The slope is plotted on the right axis and shows a slight distinction between the responses of the 400- and 30-µm thick CDC ASICs.

Figure 4.3 shows the dynamic characterization of the presented humidity sensor system. While keeping the 30-µm CDC ASIC, the off-chip/on-foil sense and reference capacitors inside the climate chamber, the chamber humidity level is varied down and then up with the maximum possible slope. As shown in Fig. 4.3a, the humidity level requires more than 30 min to cover the range from 80 to 45%. The sensor system response is plotted and it traces the dynamic changes in the humidity level of the chamber.

Since the volume of the climate-controlled chamber is relatively large (i.e. large time constant), the true response and recovery times of the implemented RH sensor can not be extracted. Therefore, the dynamic behavior of the sensor system is captured through forced adsorption by exhaled human breath, which leads to the fast saturation of the sensor. Figure 4.3b shows the measured 30-µm CDC ASIC 8-bit output code during two cycles of forced adsorption and spontaneous desorption. The rise and fall times are extracted and equals to 1.9, 0.8 s and 29, 2.5 s, respectively.

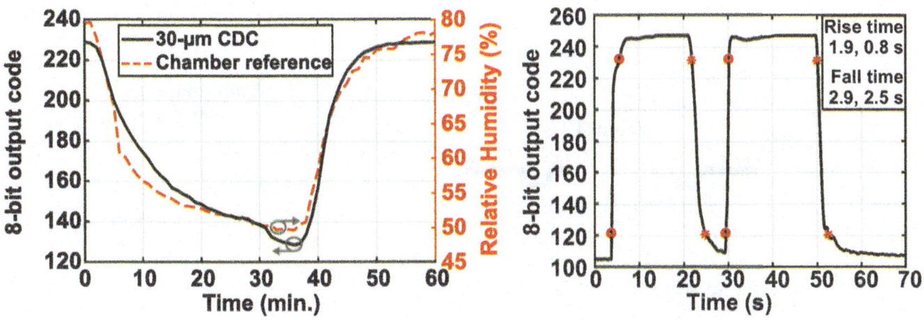

(a) (b)

Fig. 4.3 Dynamic characterization of the RH sensor readout system. **a** Measured CDC 8-bit output code and reading of the reference sensor plotted on the left and right axes versus time, while ramping the humidity level down and then up in the humidity range from 80 to 45% [13]. **b** Measured CDC 8-bit output code during forced adsorption by the exhaled human breath, which leads to sensor saturation, followed by spontaneous desorption

4.2 Strain Gauge Readout System

A smart e-skin combining multiple flexible electronic components has been demonstrated by using the CFP as a standalone HySiF [14–16]. Here, the e-skin is assembled on a robotic gripper and is used to monitor its uniaxial bending activities. Figures 4.4 and 4.5 show the cross-section and photographs of the e-skin in which UTCs are embedded in the flexible CFP package and off-chip/on-foil electronic components, such as printed strain gauges and organic addressing circuit (previously discussed in Sects. 3.3 and 3.5), are fabricated on the surface of the CFP package.

During the uniaxial bending of the robotic gripper's finger, the outer surface of the fingers experiences a non-uniform stress distribution. Therefore, three strain gauges are distributed on the upper half of the foil, where the maximum stress is present. A similar approach, presented in [17], uses only an embedded array of distributed UTCs to monitor the bending activities of the same robotic gripper. Alternatively, a single sensor readout UTC is utilized here to simultaneously readout the strain gauges. It is designed in the 0.5-μm Gate Forest™ CMOS technology [18] and is thinned down to a thickness of 20 μm for CFP embedding. In this section, an overview of the readout circuit design and electrical characterization is presented.

Architecture

Figure 4.6 shows the schematic of the employed readout circuit architecture for the off-chip/on-foil resistive strain gauges [15, 16, 19]. A resistor string digital-to-analog converter (DAC) is connected in parallel to the strain gauge Wheatstone bridge. The DAC uses the voltage drop across the bridge as a reference voltage and generates two analog voltages V_{ref1} and V_{ref2}. A single-ended differential difference amplifier (DDA) amplifies the differential bridge signal and cancels the offset voltage originating from the resistor mismatch in the Wheatstone bridge. A 10-bit SAR ADC that uses an R-2R DAC digitizes the amplified strain gauge signal at a conversion rate of 156 kS/s.

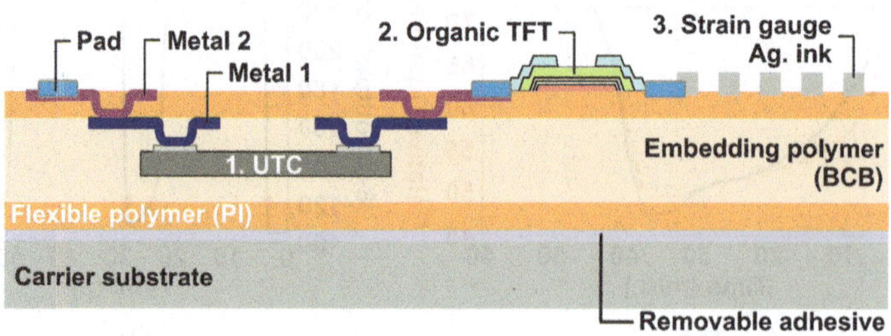

Fig. 4.4 Cross-section of one variant of the CFP process in which embedded UTC and off-chip/on-foil printed strain gauges and organic circuits are integrated into/on the polymeric substrate of the CFP flexible package [14–16]

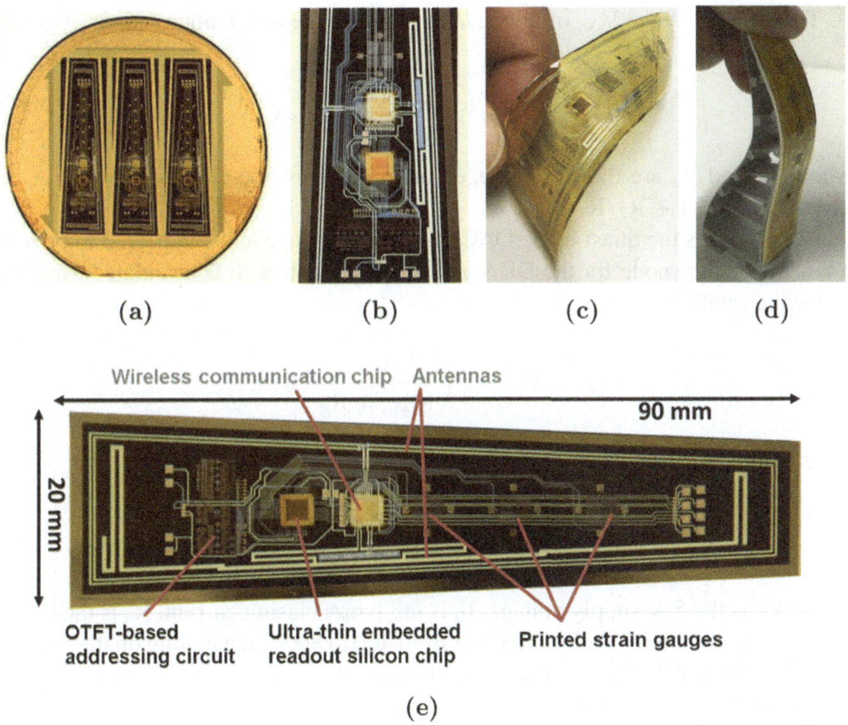

Fig. 4.5 Photographs of the implemented HySiF-based e-skin for robotic applications [16]. **a** Three HySiF strips are fabricated on each 150-mm silicon carrier wafer. **b** Zoomed-in image of the e-skin. **c** A simple mechanical release of the HySiF foil from the carrier substrate after which **d** the e-skin is assembled on the robotic gripper 3D printed plastic material using double sided foam adhesive [16]. **e** The e-skin combines two UTCs, antennas, printed strain gauges and organic circuitry

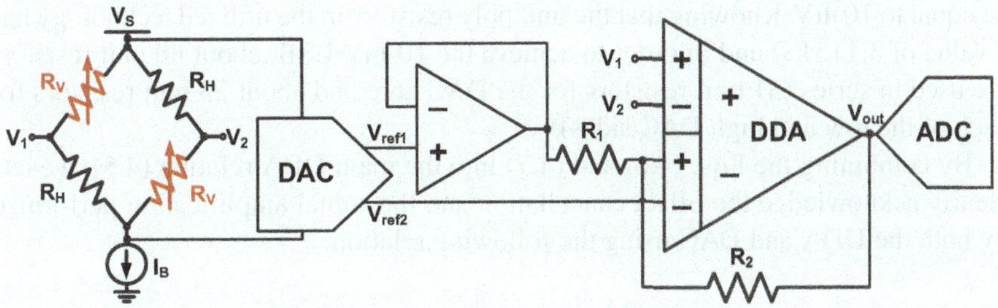

Fig. 4.6 Integrated readout system for off-chip/on-foil resistive strain gauges [15, 16]

The output voltage V_{out} of the DDA is given by the following relation:

$$V_{out} = A_{cl}\left(V_1 - V_2 + V_{ref2} - \left(1 - \frac{1}{A_{cl}}\right)V_{ref1}\right), \tag{4.5}$$

where V_1 and V_2 are the strain gauge output voltages, the closed-loop gain A_{cl} is given by $A_{cl} = 1 + R_2/R_1$.

Here, V_{ref1} is the quasi-static DAC mid-range voltage and is buffered to provide a low impedance node for the DDA gain-setting resistors. It is calculated using the following relation:

$$V_{ref1} = V_S - I_B\left(\frac{R_g}{R_{DAC} + R_g} \cdot R_{DAC}\right),$$

by assuming $R_{DAC} \gg R_g$ we conclude:

$$V_{ref1} \approx V_S - \frac{I_B}{2}(2R_g \pm \Delta R + R_{offset}), \tag{4.6}$$

where V_S is the 5-V supply voltage, I_B is the 1-mA biasing current, R_g is the bridge resistance with nominal values ranging from 400 to 800 Ω depending on the printing process. ΔR and R_{offset} are the strain dependent resistance and equivalent offset resistance values, respectively.

The DAC output voltage V_{ref2} is calculated using the following relation:

$$V_{ref2} = V_{ref1} \pm D_i \cdot LSB, \tag{4.7}$$

where D_i is the ith input digital word for the 5-bit DAC as i spans the range [0:15]. Since the bridge offset voltage is in the range of 30 mV, the DAC's LSB is designed to be equal to 10 mV. Knowing that the unit poly resistors in the utilized technology has a value of 3.115 kΩ and in order to achieve the 10-mV LSB, about 80 unit resistors are used in series (31 unit resistors for the DAC core and about 25 unit resistors for each of the low and high DAC sides).

By combining the Eqs. (4.6) and (4.7) into the main DDA relation (4.5), we can clearly acknowledge the offset cancellation and the signal amplification performed by both the DDA and DAC using the following relation:

$$V_{out} = A_{cl}\left(\frac{I_B}{2}(\Delta R + R_{offset}) \pm D_i \cdot LSB\right), \tag{4.8}$$

Measurement Results

Figure 4.7 shows the measurement setup for the e-skin in which a wafer manual probing station, oscilloscope and GUI on a laptop for data acquisition and FPGA for emulating the functionality of a customized wireline communication protocol

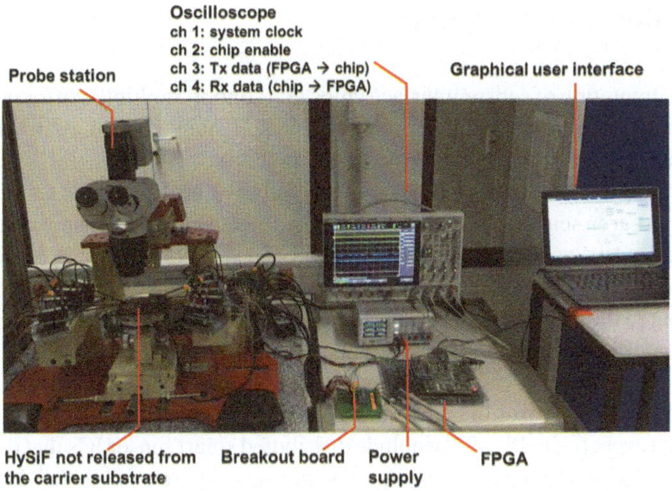

Fig. 4.7 Measurement setup for testing the e-skin (without bending)

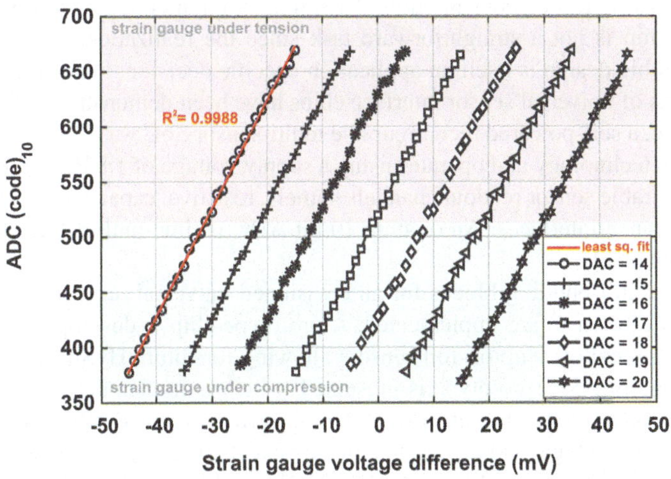

Fig. 4.8 The measurement results of the strain gauge readout channel at different offset cancellation voltages [16]

between the embedded sensor readout and another non-functional wireless communication UTC that was damaged during the CFP processing.

In order to mimic the robotic gripper bending, the bridge differential output voltage (the same as the DDA differential inputs) is swept in the range of $\pm$ 15 mV for different DAC settings. As shown in Fig. 4.8, the 10-bit ADC output code is plotted vs. the strain gauge differential voltage and good linearity is achieved. The readout system consumes 1 mA as a biasing current for the strain gauges and about 0.9 mA is consumed by the analog circuits.

4.3 MultiSense Readout Chip

The implementation of sensor interface chips that support multiple sensor types is primarily driven by the continuous miniaturization and integration of more sensors into single microsystems. It is secondarily driven, in this work in particular, by the HySiF economical approach, which minimizes the number of embedded UTCs and diversifies the large-area sensors and components. As an example, skin-inspired stretchable polymeric matrix networks integrating 8 sensors (e.g. temperature, humidity and pressure) have been fabricated on a PDMS substrate [20]. However, what is generally lacking in the literature is a universal multi-sensor interface chip that is physically compliant with flexible electronic systems.

On the contrary, universal sensor tags that are implemented using conventional PCBs and ICs (e.g. SensorTag and SmartBond manufactured by TI and dialog semiconductor, respectively [21, 22]) include distributed smart sensor ICs that do not need a centralized sensor interface circuitry. As an example, SensorTag IoT kit includes 10 smart sensors, a wireless microcontroller and a PCB antenna.

Analog circuit design is an art, which is based on tradeoffs between accuracy, speed, area and power consumption. Therefore, implementing a universal sensor interface chip is not a straightforward task since the realization of a general yet reconfigurable design is itself an application-specific decision. Previously, multiple realizations of universal sensor interface chips have been demonstrated [23–28]. As an example, a self-powered reconfigurable multi-sensor SoC was fabricated in 0.35-μm CMOS technology and operates using a supply voltage of 1.8 V [26]. It includes 4 reconfigurable sensor readout channels namely, resistive, capacitance, voltage and transimpedance interfaces. An on-chip 10-bit ADC is time-multiplexed among the different channels.

Here, several building blocks for an envisioned universal reconfigurable sensor readout (MultiSense) are implemented. A prototype chip is developed to interface with on-chip and off-chip/on-foil sensors allowing for future HySiF integration. In this section, an ASIC prototype is presented, which is implemented using a 0.18-μm CMOS technology and integrates a bandgap reference and offset-compensated capacitive and voltage readout circuits. In the next chapters, the implementation of a resolution-reconfiguration ADC is discussed. Figure 4.9 shows the layout of the first ASIC prototype, in addition to photographs of a 700- and 30-μm thick bare dies.

4.3.1 Bandgap Reference Circuit

Semiconductor technology does not directly offer any electric quantity that is independent of temperature and mechanical stress [29, 30]. For the temperature dependency, resistors fabricated using silicon or metals, having negative and positive temperature coefficients (N- and PTCs), respectively, are commonly available in IC technologies. A low TC resistor can be achieved by connecting NTC and PTC resistors

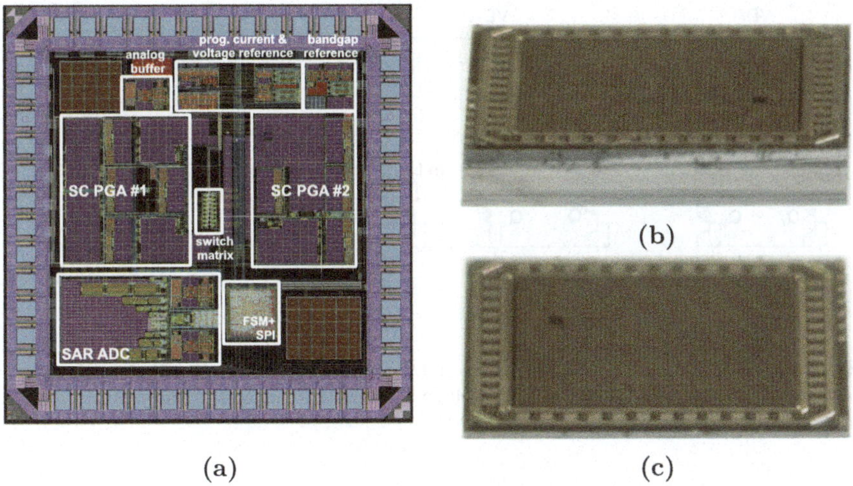

Fig. 4.9 **a** Layout and floor plan of the MultiSense readout ASIC. Photographs of **b** a 700-μm and **c** 30-μm MultiSense bare dies $1.65 \times 1.65\,\text{mm}^2$ ASICs

in series. However, the large TC mismatch and uneven sheet resistance values between NTC and PTC integrated resistors and the resistance-to-voltage conversion, have hindered the implementation of resistor-based voltage reference circuits. Alternatively, a CMOS-compatible approach, which utilizes the NTC voltage across a forward-biased diode and the PTC voltage difference between two forward-biased diodes carrying different current densities, achieves what is known as the bandgap reference circuit [31, 32].

For the mechanical stress dependency, various error sources originate during higher stress levels during chip assembly and packaging. The stress sensitivity of the reverse-saturation current traces back to the piezojunction and piezoresistive effects in diodes. Such stress-induced errors are amplified when the voltage reference is implemented in UTCs. Static stress-induced errors occur when UTCs are assembled in rigid packages (e.g. chiplets 3D integration). Dynamic stress-induced errors occur when UTCs are integrated into bendable or stretchable packaging. Static and low-frequency errors in bandgap references are trimmed, chopped corrected and shuffled (e.g. dynamic element matching (DEM) in order to cycle through different diodes and periodically select the 1 diode of the 1:n ratio in the bandgap reference) [33, 34].

Figure 4.10a shows a typical CMOS bandgap voltage reference circuit that includes two branches with equal biasing current I_B. The circuit incorporates an Operational Amplifier (op-amp) A_0 and two vertical pnp bipolar junction transistors (BJTs) Q_1 and Q_2 that are diode-connected and utilize the p-substrate as the collector terminal. The pn diode IV characteristic is described by the following equation:

$$I = I_0(e^{qV/k_B T} - 1), \tag{4.9}$$

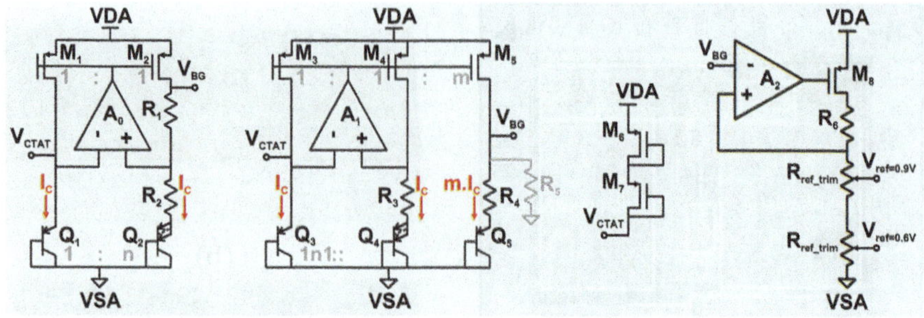

(a) (b) (c) (d)

Fig. 4.10 **a** A basic CMOS bandgap voltage reference circuit [31, 32]. **b** A low supply voltage bandgap circuit adapted from [37]. **c** Start-up branch connected to (**a**) and (**b**). **d** Bandgap regulated voltage reference circuit

where k_B is Boltzmann constant and I_0 is the reverse-saturation current, which is given by:

$$I_0 = Aqn_i^2\left(\frac{D_p}{L_pN_D} + \frac{D_n}{L_nN_A}\right), \tag{4.10}$$

where A is the diode area, n_i is the carrier concentration in undoped silicon, D_p and D_n are the diffusion coefficient for holes and electrons, N_D and N_A are the donor and acceptor dopant densities in an abrupt pn junction, L_p and L_n are the holes and electrons diffusion lengths that typically vary from a few to hundreds of μm [35]. The diffusion of the minority carriers and the electron-hole pair generation and recombination is a highly temperature dependent process. This temperature dependence is highlighted by rewriting the reverse-saturation current I_0 equation (4.10) as follows:

$$I_0 = I_0' e^{-E_g/k_BT}, \tag{4.11}$$

where E_g is the bandgap energy that itself is a function of temperature:

$$E_g(T) = E_g(0) - \frac{\alpha T^2}{\beta + T}, \tag{4.12}$$

where $E_g(0)$ is the bandgap energy at 0K, the fitting parameters α and β are $0.473\,\text{meVK}^{-1}$ and 636 K, respectively [36].

Since the base of a BJT is designed to be much narrower than the emitter, the pn junction reverse-saturation current given in Eq. (4.10) is redefined and approximately equals:

$$I_0 = k_BT\frac{A}{N_{D,A}}\mu n_i^2, \tag{4.13}$$

where A is the emitter area, μ is the mobility of the base minority charge carriers and n_i is the intrinsic concentration. Upon mechanical deformation, μ and n_i^2 vary due to the piezoresistive and piezojunction effects in silicon transistors [30]. The piezojunction effect is mainly concerned with conductivity changes due to minority carriers, which is the case in BJT transistors.

The voltage across the pn junction and hence the BJT base emitter voltage V_{BE} can be expressed as function of the collector current $I = I_C$ using the following relation:

$$V_{BE} = \frac{k_B T}{q} \ln\left(\frac{I_C}{I_0}\right). \tag{4.14}$$

The V_{BE} voltage varies nonlinearly and complimentary to absolute temperature (CTAT). This can be appreciated by calculating the temperature coefficient of V_{BE} [29], which is given by the following equation:

$$\frac{\partial V_{BE}}{\partial T} = \frac{V_{BE} - (4+m)k_B T/q - E_g/q}{T}, \tag{4.15}$$

where $m \approx -3/2$.

An intrinsically linear voltage that is proportional to absolute temperature (PTAT) can be generated using two diode-connected BJTs and is given by the following equation:

$$\Delta V_{BE} = V_{BE1} - V_{BE2} = \frac{k_B T}{q} \ln(n), \tag{4.16}$$

where n is the emitter area ratio (i.e. $A_2 = n \cdot A_1$). Using Eq. (4.11) and (4.16), the bandgap reference voltage V_{BG} of the basic circuit in Fig. 4.10a is calculated by the following equation:

$$V_{BG} = V_{CTAT} + K \cdot V_{PTAT} = V_{BE2} + \left(1 + \frac{R_1}{R_2}\right) \frac{kT}{q} \ln(n), \tag{4.17}$$

where K is the PTAT scaling factor. Note, that the op-amp offset is amplified by a factor $1 + R_1/R_2$, therefore large n and low R_1/R_2 ratio should be chosen.

Circuit Design

Figure 4.10b shows the implemented low supply bandgap reference voltage circuit, which is adapted from [37]. The output voltage V_{BG} is calculated using the following equation:

$$V_{BG} = \alpha(V_{CTAT} + K \cdot V_{PTAT}) = \frac{R_5}{R_4 + R_5}\left(V_{BE5} + m \cdot \frac{R_4}{R_3} \cdot \frac{kT}{q} \ln(n)\right), \tag{4.18}$$

where α is a scaling factor < 1 and m is the W/L ratio of M_5 to M_4 or M_3 in Fig. 4.10b. Here, the drains of M_3 and M_4 are connected to similar voltage levels ($\approx 0.6\,V$) unlike M_1 and M_2 in Fig. 4.10a (≈ 0.6 and $1.2\,V$), which leads to lower

Table 4.1 Bandgap reference circuit design parameters summary

Device	Parameter	Device	Parameter (kΩ)	Device	Parameter
Q_3, Q_5	$A_E = 10 \times 10\ \mu m^2$	**R_3**	50	**M_3, M_4**	W/L = 2.8 μm/ 1.4μm
Q_4	$A_E = 8 \times 10 \times 10\ \mu\ m^2$	**R_4**	200	**M_5**	m·W/L = 3 x 2.8 μm/1.4μm

channel length modulation and accurate PTAT current mirroring. This circuit requires a minimum supply voltage of $V_{BE} + |V_{DS}|$, as long as the op-amp supports this input and output common-mode ranges and since the nominal supply voltage here is higher, i.e. 1.8 V, the advantage of the voltage scaling resistor R_5 at the circuit output is not exploited. However, trimming the resistor R_5 directly sets the nominal V_{BG} voltage without altering its temperature coefficient. Additionally, trimming the resistor R_4 sets the PTAT voltage scaling factor (K cf. Eq. (4.18)). Figure 4.10c shows a start-up circuit, which is connected to the bandgap reference circuit in order to avoid a possible state where the two main current branches are not powered up and remain off indefinitely. Table 4.1 summarizes key design parameters for the implemented bandgap reference circuit.

Different temperature-independent voltage levels are needed in the MultiSense ASIC, namely 0.9 and 0.6 V. Figure 4.10d shows a conventional bandgap regulated voltage reference circuit in which the current source M_8 is controlled using the error signal generated from an op-amp. The output voltage of the bandgap reference circuit in addition to the op-amp and a resistive ladder are placed in a negative feedback loop, which results in temperature-independent reference voltages V_{ref}. The current source M_8 is placed in common source configuration and together with the op-amp, instability is possible especially when the op-amp itself includes multiple gain stages. As shown in Fig. 4.10d, resistor R_6 and two resistors R_{ref_trim} are used to generate the 0.9-V and 0.6-V and associated trimming voltage levels.

Measurement Results

For the bandgap reference characterization, 700- and 30-μm thick MultiSense ASICs are assembled on a customized PCB, wire-bonded and protected with plastic insulation material. Next, the ASICs are placed in a climate-controlled chamber in which the temperature is swept from −30 to 90 °C in steps of 15 °C and each step lasted for 1 h in order to reach thermal equilibrium. The ASICs are operated using a nominal supply voltage of 1.8 V and consumes currents of 4.5 and 5 μA for the op-amp and the bandgap core, respectively. The V_{CTAT} and V_{BG} voltages are buffered and multiplexed to a single output pad.

Figure 4.11 shows the measured and simulated CTAT voltages V_{CTAT} or V_{BE3} and their temperature coefficients. The plots include the typical simulation results as well as the $\pm 3\sigma$ boarders of a Monte Carlo simulation with 1000 samples, in which the beta multiplier and bandgap reference circuits' components are varied

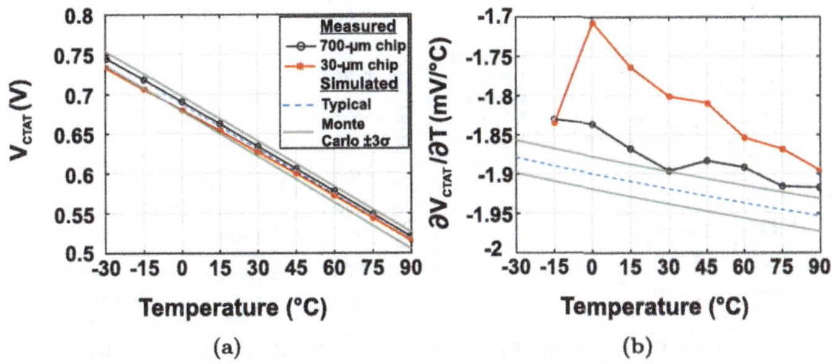

Fig. 4.11 **a** Measured and simulated CTAT voltages and **b** their temperature coefficients

using mismatch parameters from the devices' model that reflect the wafer-to-wafer and in-wafer statistical variations. Here, the Monte Carlo simulation is preferred as it is more representative of the worst case mismatch than the corners simulation.

As shown in Fig. 4.11a, the measured V_{CTAT} values for the thick and thinned ASICs are within the boundaries of the process variations in the measured temperature range. However, the extracted temperature coefficients of V_{CTAT}, plotted in Fig. 4.11b, show a similar trend of increasing values at higher temperatures, yet are slightly lower (considering the TC absolute values) than the Monte Carlo simulation $\pm 3\sigma$ boundaries. This behavior can be explained by assuming the presence of an error source with PTAT characteristics that adds up to lower the CTAT property of the V_{BE} voltage. By considering the error sources in a typical CMOS bandgap reference circuit, as presented in [33], the only error affecting V_{BE} with PTAT characteristics is the spread in the BJT reverse-saturation current ΔI_0. By rewriting Eq. (4.11) as:

$$V_{BE} = \frac{k_B T}{q} \ln\left(\frac{I_C}{I_0 + \Delta I_0}\right) \approx \frac{k_B T}{q} \ln\left(\frac{I_C}{I_0}\right) - \frac{k_B T}{q}\frac{\Delta I_0}{I_0}. \quad (4.19)$$

Assuming $\Delta I_0/I_0$ is temperature independent, it is obvious that the error in V_{BE} due to I_0 variation has a PTAT characteristics and can be removed by a single PTAT trim [33].

Aside from other PTAT error sources that might be present in the measurement setup, the spread in the reverse-saturation current ΔI_0 is higher than that modeled by the IC foundry for both ASICs. The measured TC of the CTAT voltage for the bandgap reference implemented in the 30-µm ASIC is lower compared to the 700-µm thick ASIC. Consequently, the ΔI_0 spread is higher for the 30-µm ASIC compared to the 700-µm thick ASIC. Here, one might assume that the altered wafer thickness might directly affect the reverse-saturation current since the substrate is the collector terminal of the utilized vertical pnp BJT device. However, the reverse-saturation current is independent of the collector thickness (cf. Eqs. (4.10), (4.11) and (4.13)). This directs to the higher stress levels that occur during the chips back-thinning process and thin chips assembly and packaging. The stress-induced changes in the reverse-saturation current results from combining the piezoresistive and piezojunction

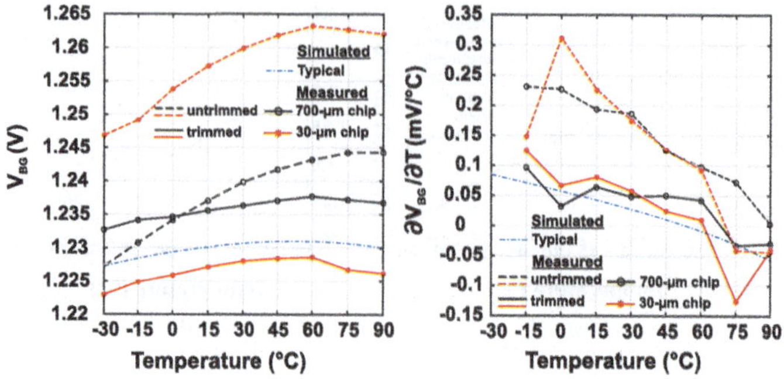

(a) (b)

Fig. 4.12 **a** Untrimmed and trimmed measured and typical simulated bandgap reference voltages and **b** corresponding temperature coefficients

effects in silicon BJTs (e.g. vertical pnp BJTs fabricated on (100) silicon wafers show highly parabolic stress-dependent reverse-saturation currents with variations up to 7% in the uniaxial stress range of ±200 MPa, [30]).

Figure 4.12 shows the untrimmed and trimmed measured and typical simulated bandgap reference voltages V_{BG} and the corresponding TCs. The plots include the measurement results of a 700-μm and 30-μm thick ASICs in addition to the typical simulation result. The measured untrimmed V_{BG} values are within the process variation boundaries, however, the untrimmed TCs of both ASICs show mostly PTAT behavior. Error sources, such as the spread in the resistors' nominal value and the mismatch between R_3 and R_4, contribute to the elevated PTAT contribution (or K factor cf. Eqs. (4.17) and (4.18)) to the V_{BG} values.

It is interesting to notice that the TC of untrimmed V_{BG} for the 700-μm and 30-μm ASICs closely match, indicating the independence of the PTAT voltage on the reverse-saturation current (c.f Eq. (4.16)). Note, that the non-PTAT error resulting from the offset voltage of the op-amp is a significant contribution to the spread of the V_{BG} values and requires offset cancellation techniques, such as chopping, that help mitigate other non-PTAT low-frequency errors. Although the output voltage range of the bandgap reference trimming resistor R_4 is overestimated during design (from 1 to 1.5 V using 4-bit RDAC), the trimmed V_{BG} values and their corresponding TCs of both ASICs are close to the typical simulated results.

From the theoretical discussion and measurement results of the bandgap reference, it is important to conclude that for low and medium precision integrated systems (below 10–12 bits), the classical bandgap reference circuits have similar performance when implemented in either thick or thin chips. A single room temperature trim corrects PTAT errors resulting from V_{BE} spread and resistor mismatch [33]. For high precision systems, additional offset cancellation and V_{BE} curvature corrections are also needed. The previously mentioned techniques are enough in systems employing thin chips with high levels of static stress.

However, for dynamically variable stress levels (e.g. in bendable or stretchable systems), the single room temperature trim of the PTAT error due to the reverse-saturation current stress dependency is no longer valid. Therefore, bandgap reference circuits should employ stress-insensitive components (e.g. silicided resistors). In addition, keeping the dominance of the PTAT voltage contribution higher than that of the CTAT voltage resulting in a general PTAT behavior that ensures stress-insensitive voltage reference (cf. Eq. (4.16)), provided that the PTAT voltage scaling factor K is also stress-insensitive.

4.3.2 *Offset-Compensated Readout Circuit*

Figure 4.13 shows a low power and compact implementation of a combined voltage and capacitive readout channel that is based on offset- and dc gain-compensated SC PGA [12]. Compared to the basic SC charge-balancing CVC implementation (cf. Fig. 4.1), an extra feedback branch is realized here by the hold capacitor C_h and associated switches. In this configuration, the output voltage V_{out} changes as little as possible between clock phases Φ_1 and Φ_2 [12]. Consequently, an OTA with a lower slew rate, thus lower power consumption, can be used. In addition, OTA matching requirements can be relaxed since the offset is compensated using the hold capacitor C_h.

Two transmission gate (TG) switches (s and $\bar{s}$ in Fig. 4.13) selects which readout channel is connected to the op-amp inverting terminal. For the capacitive readout (i.e. $s = 1$), the charge imbalance in the capacitive half bridge is related to the output voltage using the following equation:

$$V_{out} = \frac{1}{1 + \epsilon_A} \frac{C_s - C_r}{C_f} V_{DD} + V_{ref}, \tag{4.20}$$

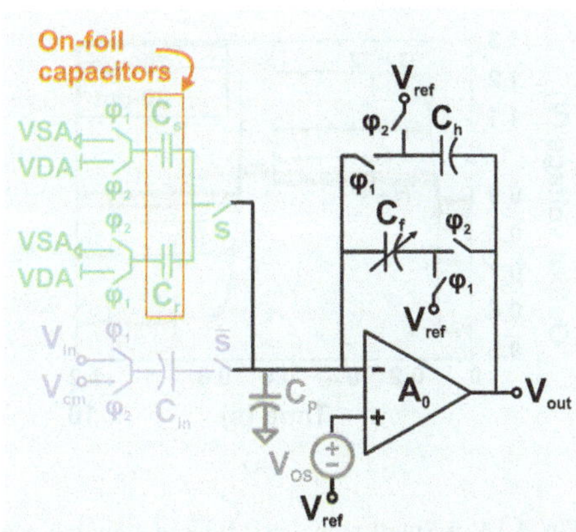

Fig. 4.13 Voltage and capacitive readout channels using offset- and dc gain-compensated SC PGA

where V_{ref} is the output common-mode voltage, here set to $V_{DD}/2$. For $C_f = C_h$, the error due to the finite dc gain A_0 of the OTA ϵ_A is given by:

$$\epsilon_A \approx \frac{1}{A_0^2}\left(1 + \frac{C_p + C_s + C_r}{C_f}\right). \tag{4.21}$$

Note, that here the error in the output voltage, for low-frequency capacitive variations, is scaled by $1/A_0^2$ compared to $1/A_0$ for the basic CVC implementation, that is previously shown in Fig. 4.1.

For single-ended voltage readout (i.e. s = 0), the output voltage is related to the input voltage V_{in} using the following equation:

$$V_{out} = \frac{1}{1 + \epsilon_A}\frac{C_{in}}{C_f}\left(V_{in}z^{-1} - V_{cm}\right) + V_{ref}, \tag{4.22}$$

where V_{cm} is the input common-mode voltage. The feedback capacitor C_f, which is implemented using a 6-bit CDAC (0.92 pF→ 29.44 pF), sets the readout closed loop gain.

Measurement Results

Figure 4.14a shows the dynamic performance of the voltage readout channel (s = 0) for small-signal inputs at different gain settings. A square wave input signal is applied with an amplitude and frequency of 20 mV and 843 Hz. The SC clock is a 1.8-Vpp signal with a frequency of 5 Hz. The output voltage shows similar behavior and no sign of stress-dependency for the 700- and 30-µm ASICs. Note, that the output voltage does not return to the common-mode voltage level every half cycle but rather stays close to the output voltage level of the previous phase. Therefore, the SC voltage readout benefits from less strengthened slew rate requirements.

Figure 4.14b shows the voltage readout channel when a large-signal is applied to the input. Here, a square wave signal is applied to the input with an amplitude and

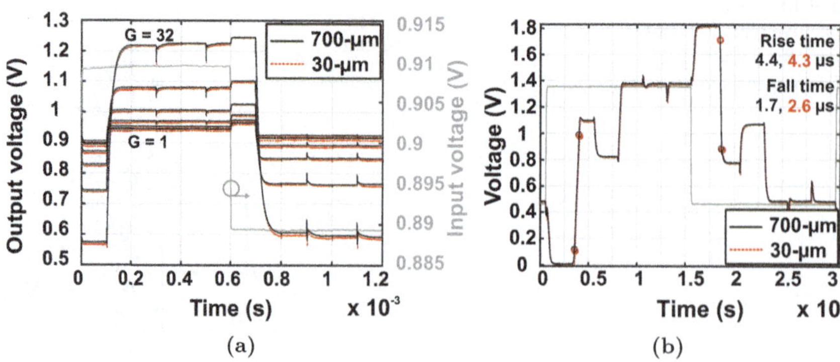

Fig. 4.14 **a** Small-signal and **b** large-signal dynamic response of the voltage readout channel

frequency of 0.9 V and 3.4 kH and a SC clock with a frequency of 20 kH. Depending on the input signal frequency, the output voltage requires more clock cycles to settle to the correct voltage level (here 2 cycles). The rise / fall times are extracted for the 700-µm and 30-µm ASICs and equal to 4.4 µs/1.7 µs and 4.3 µs/2.6 µs, respectively. When operated using a single 1.8V supply voltage, the measured current consumption of the 700-µm and 30-µm ASICs are about 135 µA and 146 µA, respectively.

The output stage in the implemented op-amp is class A with NMOS and PMOS as driver and load transistors, respectively. The constant current sourced by the PMOS load transistor (about 10 µA) sets the upper threshold of the rise time, which is independent of the input signal. This explains the similar rise times of the 700- and 30-µm ASICs. Alternatively, for the fall time, the NMOS driver transistor is able to dynamically sink current, more than the quiescent current, mainly limited by the $W/L \cdot \mu C_{ox}$ product. Here, the measured fall time of the 30-µm thick ASIC is almost twice that of the 700-µm thick ASIC, which is beyond the simulated process variations. Therefore, the only remaining effect on the fall time is the mobility variation, which in turn affects the on-resistance of the TG switches, caused by higher stress levels in the 30-µm chip.

Figure 4.15a shows the extracted transfer characteristics of the voltage readout channel at the end of Φ_2 for different clock frequencies. Gain compression occurs due to the incomplete settling of the OTA at high input amplitudes. Figure 4.15b shows transfer characteristics of the voltage readout channel when configured as a simple temperature sensor. Here, the difference between a scaled voltage of the bandgap reference and the CTAT voltage is amplified. It is interesting to observe, once more, that the TC of the thin chip is slightly lower (cf. Sect. 4.4.2).

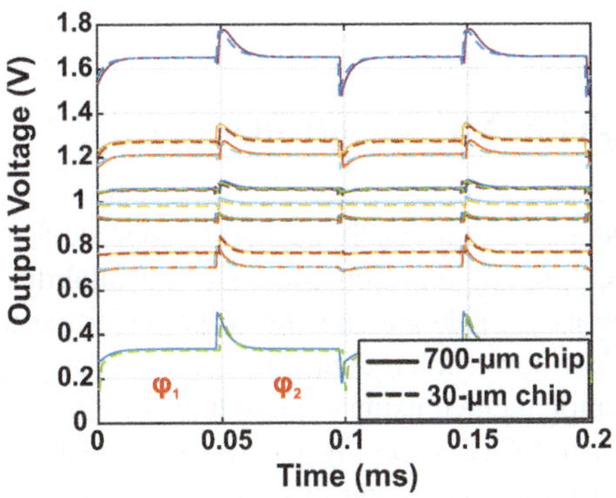

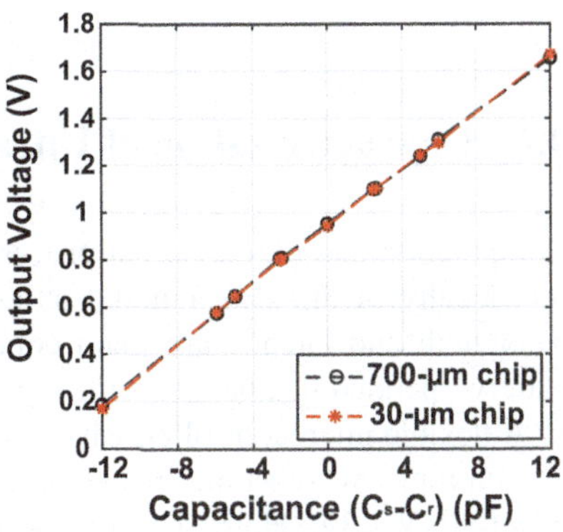

Fig. 4.15 **a** Response of the voltage readout channel at different clock frequencies to a staircase input sweep. The data points are captured at the end of the clock phase Φ_2. **b** Transfer characteristics of the voltage readout channel when configured as a simple temperature sensor. The PGA gain is set to 5, while the CTAT voltage is subtracted from the bandgap regulated voltage reference (nominal value 0.6 V)

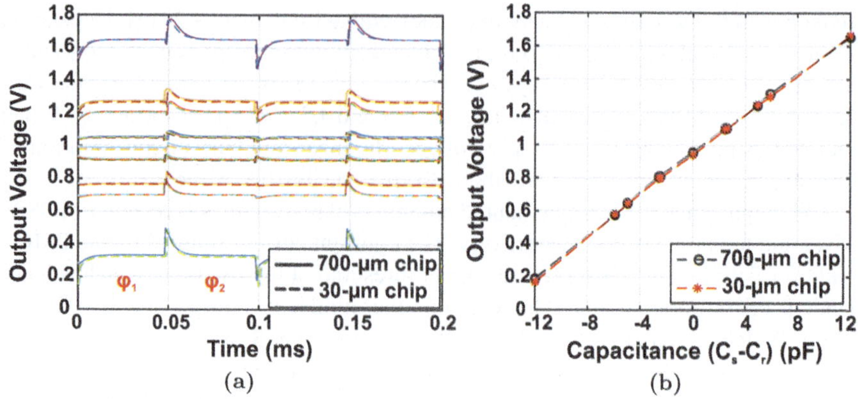

Fig. 4.16 **a** Measured response of the capacitive readout channel. **b** Extracted static data points from (**a**) at the end of the clock phase Φ_2

Figure 4.16 shows the static response of the capacitive readout channel (s = 1) for the 700-μm and 30-μm ASICs. Two discrete sense C_s and reference C_r capacitors are used to characterize the capacitive readout channel and are measured using an LCR meter, where their values equal to 48.03 pF and 47.66 pF, respectively. A maximum value of the feedback capacitor is set using the integrated CDAC and four discrete capacitors are added in parallel to either C_s or C_r to allow for the incremental increase or decrease in the output voltage. Figure 4.16b plots the capacitance change ($C_s - C_h$) vs. the extracted steady state output voltage at the end of the clock phase Φ_2. For both ASICs, a good linearity is measured and no sign of stress dependency is observed.

4.4 Characterization of Ultra-Thin Microcontroller

Computation and digital signal processing is continuously pushed to edge devices (i.e. locally in the sensor node) as user privacy and communication availability, bandwidth and latency are main concerns [38]. For achieving high-performance battery-operated flexible sensor nodes, low power microcontrollers ICs are needed.

A flexible microcontroller, which is implemented using organic thin-film transistors on a plastic substrate, is operated using a 10-V supply voltage and consumes about 90 μW [39]. However, the low operating speed and high-supply voltage are main concerns. Alternatively, silicon-based commercial microcontrollers have been thinned to less than 50 μm and embedded in polymeric fan-out packages, which were presented in various IoT-driven demonstrators, such as health monitoring, e-skin and logistics tracking [9, 16, 40, 41]. However, the performance of MCUs after the thinning and UTC embedding processes is rarely reported.

Back-thinning of bare die MCU chips is used to achieve a chip thickness of 30 μm, which is an intermediate step towards UTC embedding using the CFP fabrication

Fig. 4.17 Photographs of a 400- and b 30-µm thick low power microcontroller chips attached to probe needles for comparison

process. Figures 4.17a and 4.17b show two photographs of a 400-µm and 30-µm thick MCU chips, respectively.

The utilized MCU is known as Apollo MCU that is manufactured by the company Ambiq Micro [42]. The MCU consumes about 143 nA in deep sleep mode and 35 µA/MHz during program execution when operated using a supply voltage of 3.3 V (supply range down to 1.8 V). In this work, simple programs are designed and executed to test different MCU blocks, such as the integrated temperature sensor, 10-bit ADC and RC oscillators, before and after the chip back-thinning process [41].

4.4.1 Integrated 10-Bit ADC

Modern MCUs are fabricated using advanced deep submicron technologies (DSM) and they usually incorporate one or multiple scaling friendly ADCs, such as SAR ADCs. In the utilized MCU, a 10-bit 13-channel SAR-ADC is integrated that is able to operate at conversion rates from 115 kS/s up to 800 kS/s. The typical differential (DNL) and integral (INL) nonlinearities, from the product datasheet, equal to 1 LSB and 1.8 LSB, respectively.

For ADC characterization, the ADC is operated using a supply and reference voltages of 2 and 1.5 V. The ADC uses the internal 24-MHz RC oscillator as a clock source, which is divided by 16, in our case, for better accuracy resulting in a conversion rate of 125 kS/s. A highly linear DAC wave generator is used to generate a staircase ramp signal and the 10-bit ADC output code is captured using a Single Wire Output interface (SWO). Figure 4.18 shows the histogram linearity measurement results of the 10-bit ADC for a reference and 30-µm chips. For both chips, the DNL and INL are within 1 LSB and ±1 LSB, respectively. This proves that the chip thinning did not alter the ADC performance at the low conversion rate of 125 kS/s.

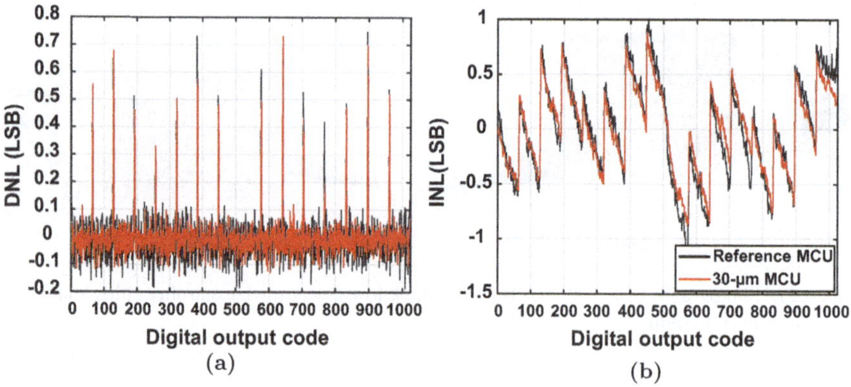

Fig. 4.18 Static histogram-based **a** DNL and **b** INL of the 10-bit SAR ADC for a reference and 30-μm thick MCUs [41]

4.4.2 *Integrated Temperature Sensor*

Compact and power-efficient temperature sensors are usually integrated into modern MCUs as they are able to quickly estimate the die temperature. For low power MCUs, the integrated temperature sensor could be used for ambient temperature estimation as chip self-heating is almost negligible, especially if the MCU is duty-cycled between long sleep and short active periods.

Here, the integrated temperature sensor, which uses a bandgap reference circuit, is characterized [41]. The 30-μm thick MCUs are assembled and wire-bonded to a ceramic package. The reference and thinned chips are placed in a climate-controlled chamber in which the temperature is varied from −20 to 100 °C in both directions and in 20 °C steps (each step lasted for 2 h until thermal equilibrium is achieved).

Figure 4.19 shows the 10-bit output code of the integrated temperature sensor plotted versus the chamber temperature for a reference and 30-μm thick MCU chips. Here, an offset error is observed between the two MCU chips that can be easily calibrated. As shown in Fig. 4.19, the sensor sensitivity is also plotted, which is calculated as the slope of the straight lines (with the unit of mV°C) using $\Delta Y/\Delta X \cdot 1.5\,V/2^{10}$. The average sensitivity in the measured temperature range slightly decreased from the nominal value of 3.6 mV°C to 3.5 mV °C. This sensitivity variation traces back to an increase in the CTAT temperature-dependent leakage current of the diode-connected devices in the bandgap reference circuit as the defects in the silicon substrate approach the device active region upon chip thinning (cf. Sect. 4.3). Note, that the offset and gain errors hint at the necessary re-calibration of ICs after the chip thinning process.

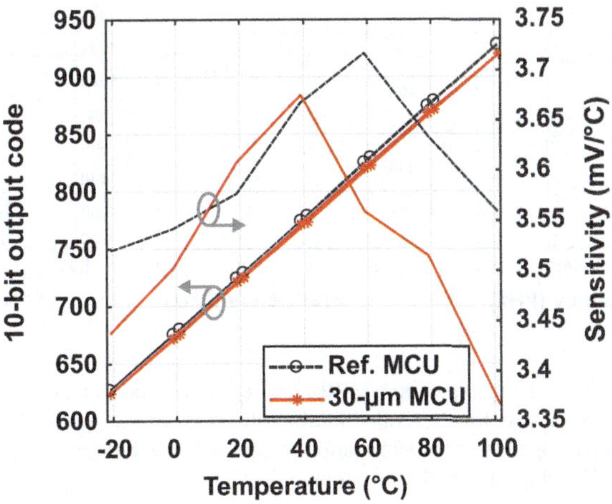

Fig. 4.19 Measured static response of the integrated temperature sensor and its sensitivity for a reference and 30-μm thick MCUs

4.4.3 RC Oscillators

The utilized MCU incorporates 1024-Hz and 24-MHz RC oscillators. The low-frequency oscillator (LFRC) is usually used for tasks with low timing accuracy and basic finite state machine operations [42], while the high-frequency oscillator (HFRC) is used when stable and accurate timing is needed (e.g. wireline communication). For characterizing the impact of the MCU chip thinning on the performance of the RC oscillators, the period jitter of the LFRC and HFRC signals is measured. The period jitter is defined as the deviation in cycle time or period of a periodic signal (e.g. clock signal) with respect to its ideal value over several randomly selected cycles [43]. Furthermore, the cycle-to-cycle jitter is defined as the time difference between the cycle time or period of each two consecutive cycles of a periodic signal.

A simple program, which activates the LFRC and HFRC oscillators, is downloaded to three MCU chips, namely a reference, 400-μm and 30-μm thick MCU chips. A high-speed oscilloscope (Tektronix 1 GS/s) is used to capture 2000 consecutive clock cycles of each MCU. Figure 4.20 shows the period jitter measurement results of the LFRC oscillator in which a histogram of the cycle frequency is plotted for the three MCUs. A slight mismatch in the LFRC oscillator frequency from the nominal frequency of 1024 H that is uncorrelated to the different MCU thickness. In general, this frequency mismatch is calibrated or trimmed using an external XTAL oscillator.

Figures 4.21a–c show the period jitter measurement results of the HFRC oscillator in which a histogram of the cycle frequency is plotted for the three MCUs. A significant degradation in the period jitter is observed for the 30-μm thick MCU. The cycle-to-cycle jitter is also plotted for the three MCU chips, as shown in Figs. 4.21d-4.21f. The extracted cycle-to-cycle jitter is within ± 2ns, ± 4 ns and ± 6 ns for the reference,

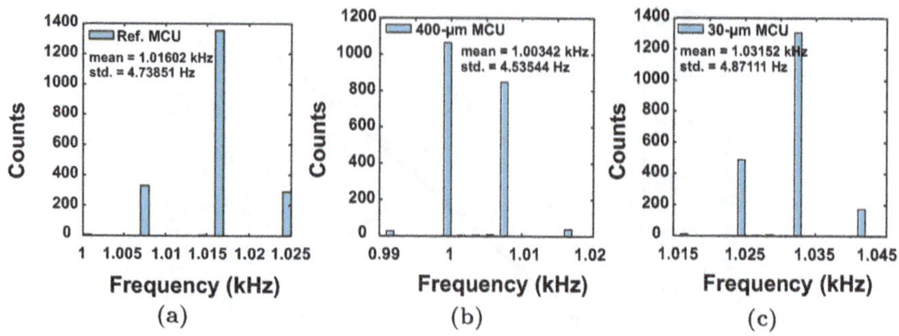

Fig. 4.20 Period jitter of the integrated LFRC oscillator for 2000 clock cycles [41, 44]. The characterization result in **a** is for the evaluation board (EVB) of the Apollo MCU, which uses a conventional BGA package. The characterization result for **b** the 400-mm thick chips and **c** is for the 30-mm thick MCU chips assembled in ceramic package

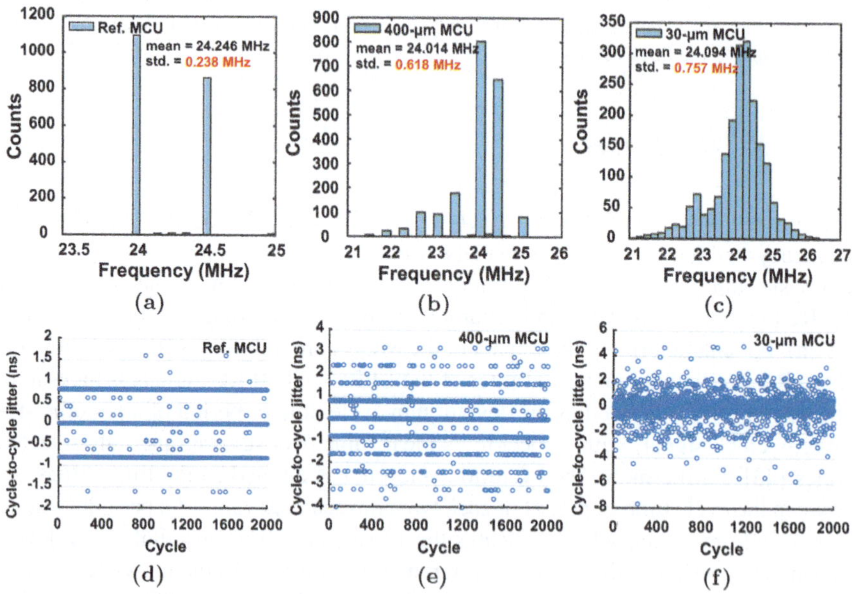

Fig. 4.21 Period and cycle-to-cycle jitter of the integrated HFRC oscillator for 2000 clock cycles [41, 44]. The characterization results in **a** and **d** are for the evaluation board (EVB) of the Apollo MCU, which uses a conventional BGA package. The characterization results for **b** and **e** the 400-mm thick chips and **c** and **f** are for the 30-mm thick MCU chips assembled in ceramic package

400-μm and 30-μm thick MCUs, respectively. Although lower resistive noise coupling in the substrate (i.e. between two p- or two n-doped regions) is expected for thinned chips as the resistive noise path is thinner than the bulk chips. For high frequencies, the capacitive noise coupling (i.e. between reverse biased p- and n-doped regions) takes effect. Here, a higher substrate digital noise capacitive coupling for the thin chips is attributed for the degradation in HFRC oscillator jitter. Traditional guard-rings are usually effective at low frequencies but such high-frequency noise coupling can be taken into consideration during UTC ASIC design using supply bounce reduction techniques, such as those presented in [45]. Note, that noise coupling is minimized as long as the circuits are placed at a certain distance, however, this is ineffective in those circuits with densely packed devices, such as oscillators, phase-locked loops (PLLs) and image sensor pixels [45].

References

1. Ghanbari MM, Tsai JM, Nirmalathas A, Muller R, Gambini S (2017) An energy-efficient miniaturized intracranial pressure monitoring system. IEEE J Solid-State Circ 52(3):720–734
2. Oh S, Jung W, Yang K, Blaauw D, Sylvester D (2014) 15.4b incremental sigma-delta capacitance-to-digital converter with zoom-in 9b asynchronous SAR. In: 2014 symposium on VLSI circuits digest of technical papers, June 2014, pp 1–2
3. Tan Z, Daamen R, Humbert A, Souri K, Chae Y, Ponomarev YV, Pertijs MAP (2011) A 1.8V 11μW CMOS smart humidity sensor for RFID sensing applications. In: IEEE Asian solid-state circuits conference 2011, Nov 2011, pp 105–108
4. Yang R, Pertijs MAP, Nihtianov S (2017) A precision capacitance-to-digital converter with 16.7-bit ENOB and 7.5-ppm/^{o}C thermal drift. IEEE J Solid-State Circ 52(11):3018–3031
5. Tan Z, Shalmany SH, Meijer GCM, Pertijs MAP (2012) An energy-efficient 15-bit capacitive-sensor interface based on period modulation. IEEE J Solid-State Circ 47(7):1703–1711
6. Jung W, Jeong S, Oh S, Sylvester D, Blaauw D (2015) 27.6 a 0.7pF-to-10nF fully digital capacitance-to-digital converter using iterative delay-chain discharge. In: 2015 IEEE international solid-state circuits conference - (ISSCC) digest of technical papers, Feb 2015, pp 1–3
7. Ha H, Sylvester D, Blaauw D, Sim J (2014) 12.6 a 160nW 63.9fJ/conversion-step capacitance-to-digital converter for ultra-low-power wireless sensor nodes. In: 2014 IEEE international solid-state circuits conference digest of technical papers (ISSCC), Feb 2014, pp 220–221
8. Oh S, Lee Y, Wang J, Foo Z, Kim Y, Blaauw D, Sylvester D (2014) Dual-slope capacitance to digital converter integrated in an implantable pressure sensing system. In: ESSCIRC 2014—40th European solid state circuits conference (ESSCIRC), Sep 2014, pp 295–298
9. Elsobky M, Albrecht B, Richter H, Burghartz JN, Ganter P, Szendrei K, Lotsch BV (2017) Ultra-thin relative humidity sensors for hybrid system-in-foil applications. In: 2017 IEEE sensors, Oct 2017, pp 1–3
10. Elsobky M, Alavi G, Albrecht B, Deuble T, Harendt C, Richter H, Yu Z, Burghartz J (2018) Ultra-thin sensor systems integrating silicon chips with on-foil passive and active components. In: Proceedings, vol 2, p 748
11. Warkentin K, Feucht G, Richter H (2004) Entwicklung eines mixed-signal asics mit dem ims gate-forest dargestellt am beispiel einer kapazitäts-messschaltung mit digitalem ausgang; development of a mixed-signal asics with ims gate-forest demonstrated on a capacitance readout circuit with digital output. In: MPC-workshop 2004—31st multi project chip group (MPC), Feb 2004, pp 21–26

12. Unbehauen R, Cichocki A (1989) MOS switched-capacitor and continuous-time integrated circuits and systems. Springer, Berlin Heidelberg
13. Elsobky M, Albrecht B, Ganter P, Szendrei K, Deuble T, Harendt C, Lotsch B, Burghartz JN (2017) Relative humidity sensors for system-in-foil applications. In: MikroSystemTechnik 2017; congress, Oct 2017, pp 1–4
14. Hassan M, Keck J, Klauk H, Kostelnik J, Mahsereci Y, Sailer S, Schreivogel A, Zaki T, Burghartz JN (2015) Combining organic and printed electronics in hybrid system in foil (HySiF) based smart skin for robotic applications. In: 2015 European microelectronics packaging conference (EMPC), Sep 2015, pp 1–6
15. Elsobky M, Mahsereci Y, Keck J, Richter H, Burghartz J (2017) Design of a CMOS readout circuit on ultra-thin flexible silicon chip for printed strain gauges. Adv Radio Sci 15:123–130
16. Elsobky M, Mahsereci Y, Yu Z, Richter H, Burghartz JN, Keck J, Klauk H, Zschieschang U (2018) Ultra-thin smart electronic skin based on hybrid system-in-foil concept combining three flexible electronics technologies. Electron Lett 54(6):338–340
17. Mahsereci Y, Sailer S, Richter H, Burghartz J (2015) 16.1 an ultra-thin flexible CMOS stress sensor demonstrated on an adaptive robotic gripper. In: 2015 IEEE international solid-state circuits conference—(ISSCC) digest of technical papers, Feb 2015, pp 1–3
18. (2019) IMS CHIPS homepage. [Online]. Available: https://ims-chips.de/
19. Elsobky M (2015) Design of a flexible smart skin for robotic application. Master's thesis, University of Ulm
20. Hua Q, Sun J, Liu H, Bao R, Yu R, Pan C, Wang Z (2018) Skin-inspired highly stretchable and conformable matrix networks for multifunctional sensing. Nat Commun 9:12
21. (2019) Texas Instruments SimpleLink multistandard CC2650 SensorTag kit reference design (Online). Available: https://www.ti.com/tool/TIDC-CC2650STK-SENSORTAG
22. (2019) Dialog semiconductor IoT multi sensor developement kits (Online). Available: https://www.dialog-semiconductor.com/products/da14585-iot-multi-sensor-development-kit
23. Zhang J, Zhang K, Wang Z, Mason A (2002) A universal micro-sensor interface chip with network communication bus and highly programmable sensor readout. In: The 2002 45th midwest symposium on circuits and systems, 2002. MWSCAS-2002, vol. 2, Aug 2002, p II
24. Zhang J, Zhou J, Mason A (2007) Highly adaptive transducer interface circuit for multiparameter microsystems. IEEE Trans Circ Syst I Regul Pap 54(1):167–178
25. Lu JH, Inerowicz M, Joo S, Kwon J, Jung B (2011) A low-power, wide-dynamic-range semi-digital universal sensor readout circuit using pulsewidth modulation. IEEE Sens J 11(5):1134–1144
26. Huang Y, Tzeng T, Lin T, Huang C, Yen P, Kuo P, Lin C, Lu S (2014) A self-powered CMOS reconfigurable multi-sensor SoC for biomedical applications. IEEE J Solid-State Circ 49(4):851–866
27. Xu J, Konijnenburg M, Ha H, van Wegberg R, Lukita B, Asl SZ, Van Hoof C, an Helleputte N (2017) A 36μW reconfigurable analog front-end IC for multimodal vital signs monitoring. In: 2017 symposium on VLSI circuits, pp C170–C171
28. Xin H, Andraud M, Baltus P, Cantatore E, Harpe P (2019) A 0.34-571nW all-dynamic versatile sensor interface for temperature, capacitance, and resistance sensing. In: ESSCIRC 2019—IEEE 45th European solid state circuits conference (ESSCIRC), pp 161–164
29. Razavi B (2016) The bandgap reference [a circuit for all seasons. IEEE Solid-State Circ Mag 8(3):9–12
30. Creemer J, French P (2004) The saturation current of silicon bipolar transistors at moderate stress levels and its relation to the energy-band structure. J Appl Phys 96:4530–4538
31. Widlar R (1965) Some circuit design techniques for linear integrated circuits. IEEE Trans Circ Theory 12(4):586–590
32. Widlar RJ (1971) New developments in IC voltage regulators. IEEE J Solid-State Circ 6(1):2–7
33. Ge G, Zhang C, Hoogzaad G, Makinwa KAA (2011) A single-trim cmos bandgap reference with a 3σ inaccuracy of $\pm$0.15% from -40°C to 125°C. IEEE J Solid-State Circ 46(11):2693–2701

34. Herbst S (2011) A low-noise bandgap voltage reference employing dynamic element matching. Ph.D. dissertation, Massachusetts Institute of Technology
35. Hu CC (2010) Modern semiconductor devices for integrated circuits. Pearson
36. Zeghbroeck BV (2011) Principles of semiconductor devices and heterojunctions. Prentice Hall, Hoboken
37. Neuteboom H, Kup BMJ, Janssens M (1997) A DSP-based hearing instrument IC. IEEE J Solid-State Circ 32(11):1790–1806
38. Bankman D, Yang L, Moons B, Verhelst M, Murmann B (2019) An always-on 3.8 μJ/86 with all memory on chip in 28-nm CMOS. IEEE J Solid-State Circ 54(1):158–172
39. Myny K, van Veenendaal E, Gelinck GH, Genoe J, Dehaene W, Heremans P (2012) An 8-bit, 40-instructions-per-second organic microprocessor on plastic foil. IEEE J Solid-State Circ 47(1):284–291
40. Sterken T, Vanfleteren J, Torfs T, de Beeck MO, Bossuyt F, Van Hoof C (2011) Ultra-thin chip package (UTCP) and stretchable circuit technologies for wearable ECG system. In: 2011 annual international conference of the IEEE engineering in medicine and biology society, Aug 2011, pp 6886–6889
41. Elsobky M, Ottaviani A, Alomari M, Yu Z, Deuble T, Burghartz JN (2019) Characterization of thin-film temperature sensors and ultra-thin chips for HySiF integration. In: 2019 IEEE international conference on flexible and printable sensors and systems (FLEPS), July 2019, pp 1–3
42. (2019) AmbiqMicro homepage (Online). Available: https://www.ambiqmicro.com/mcu/
43. IEEE standard for transitions, pulses, and related waveforms. IEEE Std 181-2011 (Revision of IEEE Std 181-2003), pp 1–71
44. Elsobky M, Deuble T, Ferwana S, Albrecht B, Harendt C, Ottaviani A, Alomari M, Burghartz JN (2020) Characterization of on-foil sensors and ultra-thin chips for HySiF integration. IEEE Sens J: 1
45. Nagata M, Nagai J, Hijikata K, Morie T, Iwata A (2001) Physical design guides for substrate noise reduction in CMOS digital circuits. IEEE J Solid-State Circ 36(3):539–549

Chapter 5
Reconfigurable Analog-to-Digital Converter for HySiF: Part I

One of the main goals of this work is the implementation of a power-efficient, sample rate and resolution reconfigurable ADC with state-of-the-art performance. The ADC design is targeting smart electronic systems that integrate multiple heterogeneous sensors. By exploiting the back-thinning of the ADC chips as a post-processing step, the presented ADC evolves into a HySiF-compatible UTC, which is readily available for integration into flexible electronic systems. The presentation of the proposed ADC is divided in this work into two chapters. This first chapter deals with the literature review and relevant theory, while the next chapter presents the ADC's circuit-level implementation and measurement results.

This chapter starts with briefly investigating the general ADC trade-offs, specifying the expected performance and outlining the top-down design methodology. Next, reconfigurable ADCs, as well as ADCs that have been implemented on flexible substrates, are reviewed. Furthermore, a short discussion of $\Delta\Sigma$ modulators promotes the in-depth analysis of incremental $\Delta\Sigma$ (I-$\Delta\Sigma$) ADCs. Finally, the proposed architecture for a power-efficient reconfigurable ADC is presented.

5.1 Motivation, Target and Methodology

The ADC design space can be navigated using standard performance metrics, such as Signal-to-Noise+Distortion Ratio (SNDR), Schreier Figure-of-Merit (FoM_S), Walden Figure-of-Merit (FoM_W), etc. Fig. 5.1 plots the energy consumption of recently published ADCs versus SNDR [1]. Selected I-$\Delta\Sigma$ ADCs are highlighted and a trendline is extracted in order to convey an impression about the reconfigurability potential of I-$\Delta\Sigma$ ADCs. Note, that reconfigurable SAR ADCs dominate the low to medium resolution range that is below SNDR of 60 dB [2, 3]. Besides, a reconfigurable time-to-digital converter (TDC) has been recently reported to span

M. Elsobky, *Ultra-Thin Sensors and Data Conversion Techniques for Hybrid System-in-Foil*, Recognizing Outstanding Ph.D. Research,
https://doi.org/10.1007/978-3-030-97726-9_5

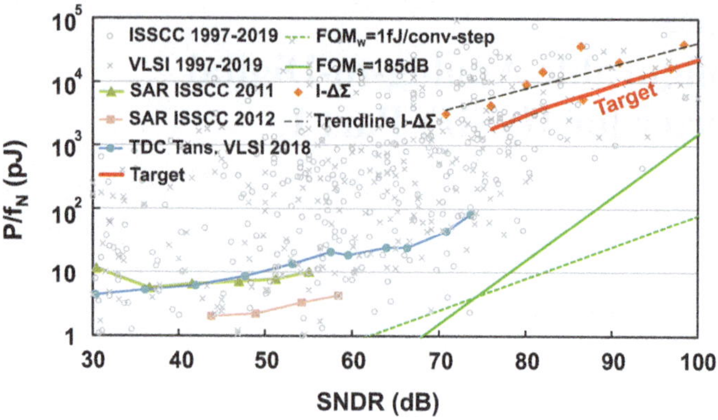

Fig. 5.1 Energy consumption versus SNDR of recently published ADCs adapted from [1]

Table 5.1 Summary of the specifications for the proposed ADC

Resolution	12–16 b
Nyquist rate	40 kS/s (16 b)
Technology	XFAB XH018 180-nm CMOS
Supply voltage	3.3 V
Power	≤1 mW
Architecture	I-ΔΣ, 3rd order CIFF
Chip thickness	700- to 30-μm

the 5–13-b resolution range with SNDR from 30 to 74 dB [4]. Consequently, it is now evident that there is an innovation opportunity for medium to high reconfiguration resolutions beyond SNDR of 80 dB in the field of Nyquist-rate ADCs.

In this work, this opportunity is exploited by proposing a power-efficient scalable SC I-ΔΣ ADC with wide reconfiguration range between 12 and 16 b. The I-ΔΣ architecture is chosen in order to deliver high resolution when digitizing multiple on-chip and on-foil sensors. Additionally, no calibration is required either in the analog or digital domains resulting in reduced system complexity. Besides, ADCs employing feedback systems are robust and more tolerant of nonidealities of the underlying building blocks [5]. Consequently, I-ΔΣ are a reasonable candidate for implementing a reliable ADC that is less sensitive to process- and bending-induced stresses.

Table 5.1 lists the initial specifications for the proposed ADC. The ADC's conversion speed is set in the kilohertz range, which targets sensor readout applications, such as temperature and humidity sensor systems. Besides, for the later HySiF integration, the ADC is tested when the bulk chip is back-thinned to thicknesses of 100 and 30 μm.

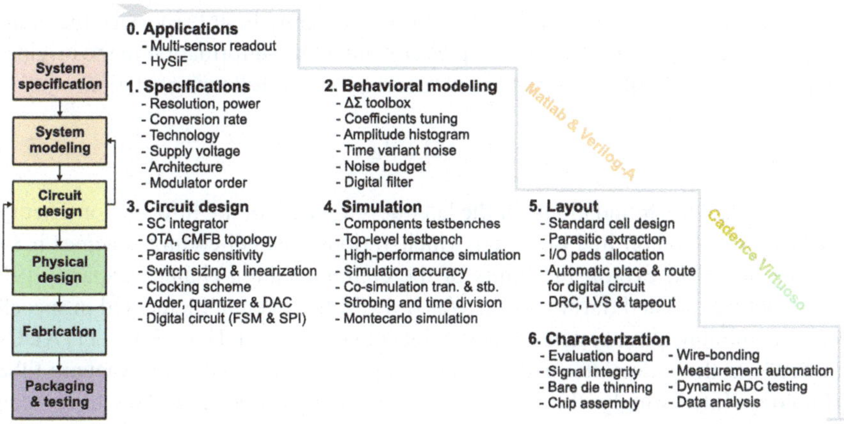

Fig. 5.2 ADC design tasks and methodology developed in this work

Figure 5.2 shows the design tasks and top-down methodology which are developed for implementing the presented ADC. Motivated by the targeted applications and after an in-depth literature review, the ADC design specifications have been formulated. Next, $\Delta\Sigma$ toolbox is utilized for the initial estimation for the coefficients of the discrete-time (DT) $\Delta\Sigma$ modulator [6]. Furthermore, Verilog-A models for analog blocks, such as OTA and switches, are used to transform the mathematical representation of $\Delta\Sigma$ modulator into circuit-level models. Additionally, the ADC circuit is divided into unit blocks to facilitate the later floor-planning and layout. Transistor-level design and simulation are performed based on the device models for the XH018 CMOS technology provided by the company X-FAB Silicon Foundries.

The default settings for the Spectre simulation platform from Cadence Virtuoso is altered to guarantee accurate simulation results. Moreover, multithreading using the accelerated parallel simulator (APS) is enabled to accelerate simulation time without losing the accuracy of the analog simulator. Similarly, simulation testbenches use co-simulation of transient and stability analysis to ensure proper operation of each building block. After tapeout, an evaluation board is designed on which the ADC chips with different thicknesses are assembled. Finally, the dynamic ADC testing method is used to evaluate the performance of the presented ADC.

5.2 Analog-to-Digital Converters

By the late 19th century, the field of communications has been the main reason for the development of electronic data converters [7]. Aiming at finding the maximum signaling rate that could be used over a channel with a given bandwidth, Harry Nyquist formulated the well-known Nyquist sampling theorem in 1920. While studying pulse code modulation (PCM), he concluded that when a signal with bandwidth

f_B is sampled instantaneously at regular intervals at a rate f_N at least twice the highest significant frequency, then the samples contain all the information in the original signal [7, 8]. The Nyquist-theorem is summarized in the following relation:

$$f_N \geq 2 \times f_B. \quad (5.1)$$

This has laid the foundations for the later development of the Information Theory by Claude E. Shannon in 1948, after which various channel-coding schemes have been developed to exploit the limits of the communication channel capacity [9]. Shortly before that and during World War II, PCM has been implemented in speech secrecy communication systems, where 5-bit counting [10] and later SAR [11] ADCs and DACs at the transmitter and receiver have been integrated using vacuum tube technology. Interestingly, the main ADC architectures used nowadays have been invented in the early 20th century [7].

Figure 5.3 shows the general resolution versus signal bandwidth capabilities of main data converters architectures. Flash ADCs are among the fastest ADC architectures but with a limited resolution and large size. Pipeline ADCs are used in medium- to high-speed communication and imaging systems. They employ a cascade of multiplying stages, which have a pipeline latency disadvantage depending on the number of stages. But once filled, pipeline ADCs optimally utilize a serial data link. SAR ADCs consist of a hybrid combination of digital logic, comparator and analog DAC. Their digital circuit implements search algorithms that have benefited from scaled DSM technologies. ADCs that incorporate a $\Delta\Sigma$ modulator (e.g. freely-running, I-$\Delta\Sigma$ and noise-shaping (NS) SAR ADCs) unlocks higher resolution capabilities by the virtue of oversampling and in-band noise-shaping. Dual-slope ADCs are among the integrating ADCs category in which the input signal charges and/or discharges a capacitor at a constant rate. Meanwhile, a counter calculates the ramp-up and/or ramp-down times. Therefore, a direct trade-off between resolution and speed arises.

Nowadays, hybrid and digitally-assisted ADC architectures are increasingly used as they efficiently complement the conventional ones. Hybrid ADCs are not limited to the topology-level combination of single conventional ADCs but rather are extended to the combination of different signal processing techniques, circuit-level optimization methods and advanced data conversion algorithms [12]. As an example, the linearity of SAR ADCs is extended beyond 100 dB by employing the DAC mismatch error shaping (MES) technique [13]. Besides, in order to mitigate dynamic range losses due to MES, the first three most significant bits (MSB) are linearized using data weighted averaging (DWA) and are resolved using Flash ADC, thus saving switching power. Finally, an oversampling NS $\Delta\Sigma$ modulator samples the residue voltage of the SAR least significant bit (LSB) and shapes the comparator noise and quantization noise.

While incorporating Flash, SAR and $\Delta\Sigma$ together with DEM, MES and dithering, the aforementioned ADC's reconfigurability results in a 9.5- and 16.5-b resolutions for Nyquist operation up to conversion rate of 5 MS/s and an oversampling operation at 2 kS/s, respectively. In this context, hardware reconfiguration targeting an

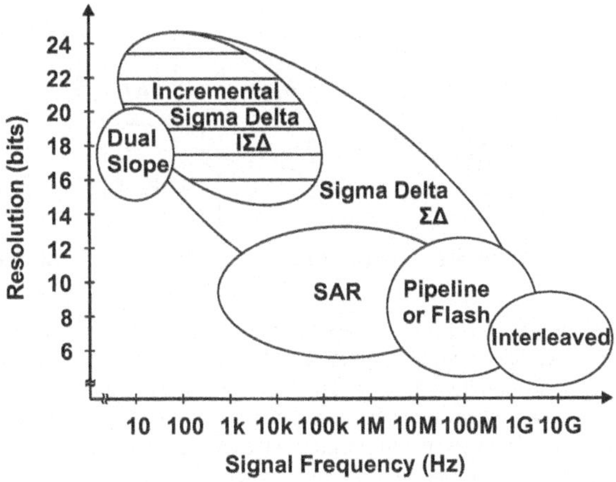

Fig. 5.3 Resolution versus signal bandwidth capabilities of main data converters architectures

optimized data conversion is a key ingredient of hybrid ADCs [12]. In the following section, state-of-the-art resolution and speed reconfigurable Nyquist-rate ADCs are reviewed. Next, selected examples of mechanically flexible ADCs and their corresponding TFT technologies are highlighted.

5.2.1 Reconfigurable ADCs

Power-efficient sensor SoCs are essential components for IoT applications. Such SoCs frequently integrate multiple heterogeneous sensors in/on the same package or die. Here, each sensor can require a different dynamic range (DR) and signal bandwidth in the subsequent ADC. As an example, an environmental sensor can switch from its standard low-resolution mode to a higher resolution mode upon a sudden change in its measured data that requires more accurate monitoring for further investigations.

Figure 5.4 shows typical multi-sensor readout channels where a single ADC is shared among different channels. In a system using a fixed, non-reconfigurable ADC, the ADC has to comply with the highest SNR and speed requirements, resulting in poor power efficiency when digitizing the lower resolution sensor data. To tackle this problem, resolution reconfigurable Nyquist-rate ADCs have been proposed. They aim at delivering a better end-to-end power efficiency when a single ADC is time-multiplexed between different sensors and/or when a sensor with dynamically reconfigurable resolution modes has to be digitized.

For low resolutions, a 3–7 b hybrid two-step and Flash ADC has been reported, which operates from 1.5 to 4 GS/s conversion rates [14]. It requires digital calibration and a complex clocking scheme. For low to medium resolutions, resolution reconfigurable SAR ADCs, which use a combination of a scalable CDAC and supply voltage [2] or a scalable CDAC and comparator [3], have been proposed. They can

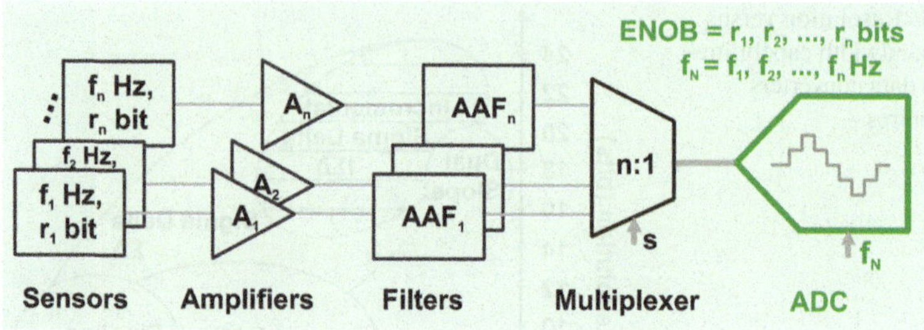

Fig. 5.4 Typical multi-sensor readout channels where a single ADC is shared among n channels. Each sensor delivers a signal accuracy and bandwidth of r_i bits and f_i Hz, where i is the index of the corresponding channel. The usual practice of using a fixed resolution and speed ADC results in a poor power efficiency during the operation of the lower resolution and/or lower speed sensor channels. Therefore, resolution and speed-reconfigurable ADCs aim at bringing about an optimized power efficiency for all readout channels

span the 5–10 b resolution range at few MS/s conversion rates while maintaining a low FoM_W of less than 10 fJ/conv.step.

An attempt to extend the resolution reconfigurable range of SAR ADCs to 14 b is achieved by employing start-up calibration, post-corrections and dithering [15]. However, increased system complexity and the resulting large area of the digital circuitry are main concerns. In [16], a first-order NS SAR ADC is used for the 8–10 b resolutions and is reconfigured as an adder and quantizer within a third-order $\Delta\Sigma$ in order to achieve resolutions up to 16 b. However, a freely-running $\Delta\Sigma$ modulator hinders the Nyquist-rate A/D conversion. In conclusion, while the SAR-based approach is suitable for the low to medium resolution design, it cannot be easily or efficiently extended towards higher resolution.

For medium resolutions, multiple power scalable bandwidth and/or resolution reconfigurable pipeline ADCs have been reported. As an example, parallel OTA scaling has been proposed, which enables a near-constant FoM_W for a fixed 12 ADC operating at reconfigurable conversion rates from 20 to 40 MS/s [17]. However, pipeline ADCs are limited to the medium resolution range owing to their topological dependency on the absolute process and design parameters, such as capacitors matching and OTA open-loop gain and settling errors.

Aside from the voltage-domain ADCs, a TDC, which operates at 500 MS/s-5 GS/s and 5–13 b conversion rate and resolution reconfigurable ranges, has been proposed [4]. The TDC is configured either as SAR-TDC or Flash-TDC for high resolution or high conversion rates, respectively. Limited by its voltage-controlled oscillator (VCO) nonlinearity, a foreground digital calibration is required. Table 5.2 compares selected Nyquist-rate ADCs reflecting the current state of the art in the field of reconfigurable ADCs.

Table 5.2 Comparison of selected Nyquist-rate ADC topologies

Topology	SAR	Pipeline	I ΔΣ	TDC
Resolution	🟫🟦⬜	⬜🟦⬜	⬜🟦🟩	🟫🟦⬜
Speed	🟫🟦🟩	⬜🟦⬜	🟫🟦⬜	⬜🟦🟩
Technology scaling	☺	☹	☹	☺
Calibration free	☹	☹	☺	☹
Tolerance to process- and stress-induced errors	😐	☹	☺	☺
Reconfigurable hardware utilization	😐	☹	☹	☺

5.2.2 *Ultra-Thin Flexible ADCs*

The next generation of mechanically flexible RFID modules calls for miniaturization (i.e. more thickness reduction), higher computational power and more integrated sensors and sensor readout channels while keeping minimal power consumption. ADCs are at the heart of every smart sensor system and notably required for achieving high performance flexible electronic systems. Although ADCs integrated into ultra-thin chips represent the natural extension for the silicon-based semiconductor technology to the flexible electronics world, ADCs fabricated using other technologies, notably organic and inorganic metal-oxide semiconductors, are on the rise. These technologies allow for large-area coverage and are inherently flexible and simply processed in a room-temperature environment.

Figure 5.5a shows a photograph of the Silicon-on-Polymer package developed by the company American Semiconductor in which the industry's first bendable ADC is embedded [18]. The 8-channel 8-bit 100-kS/s SAR ADC is fabricated using a 180-nm CMOS technology and its die size is about $2.5 \times 2.5\,\text{mm}^2$. The so-called FleX-ICs are essentially UTCs with thickness of 10–25 μm and are embedded in 35–50-μm thick polymeric packages using spin-on polymers.

Alternatively, Fig. 5.5b shows a photograph of the 6-b SAR ADC fabricated on PI substrate using unipolar dual-gate metal-oxide (InGaZnO) TFTs [19]. The ADC occupies an area of $27.5\,\text{mm}^2$ and consumes about 73 μW when operated using a supply voltage of 15 V at a conversion rate of 26.67 S/s. Other examples include $\Delta\Sigma$-based ADCs fabricated on polymeric foils utilizing OTFTs [5] and InGaZnO TFTs [20] achieving FoM_W of 3.45 and 0.039 μJ/conv.-step, respectively. Table 5.3 compares different TFT technologies that are used to fabricate flexible ADCs, namely single-crystalline UTCs, poly-crystalline silicon (poly-Si), amorphous silicon (a-Si), OTFTs and metal-oxide TFTs (e.g. InGaZno).

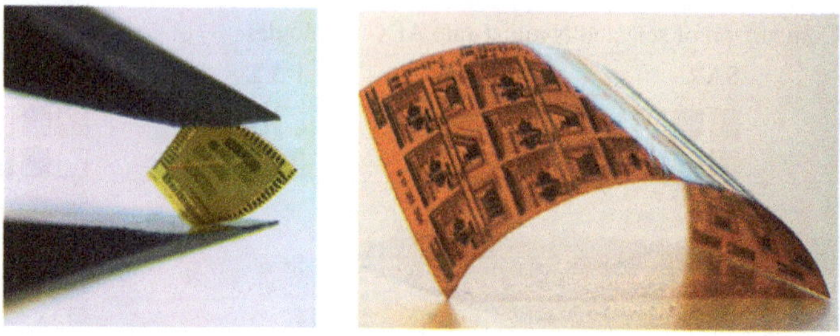

(a) (b)

Fig. 5.5 **a** Photograph of 50-μm thick Silicon-on-Polymer package with an embedded 10–25-μm FleX-ICs developed by the company American Semiconductor. Reproduced from [18] with permission from American Semiconductor. **b** Photograph of 6-b SAR ADC fabricated on PI substrate using metal-oxide TFT technology. Reproduced from [19] with permission from IEEE

Table 5.3 Comparison of different TFT technologies used to fabricate flexible ADCs

Technology	Si-UTC	Poly-Si	a-Si	OTFT	Metal-oxide
Resolution	■■■	■■□	■□□	■□□	■□□
Speed	■■■	■■□	■■□	■□□	■□□
CMOS availability	☺	☺	☹	😐	☹
Room-temperature processing	☹	☺	☺	☺	☺
Large-area coverage Roll-to-roll	☹	😐	☺	☺	☺

5.3 ΔΣ Data Converters

Since both freely-running and I-ΔΣ ADCs have intrinsically similar structures, it is rather intuitive to briefly discuss the theory behind ΔΣ data converters. Figure 5.6 shows a block diagram of a ΔΣ ADC [21, 22]. The antialiasing filter (AAF) limits the bandwidth of the analog input signal to half the Nyquist sampling frequency $f_{N/2}$ [22]. Therefore, higher frequency components are blocked and do not alias back during the sampling process. The sample-and-hold process (S/H) samples the continuous input signal u(t) and holds it constant u(n) for the duration of a single A/D conversion. The sampling process is repeated at a certain rate which correlates with the ADC's conversion rate and must not violet Nyquist theorem (cf. (5.1)) for proper reconstruction of the signal in the digital domain. The ΔΣ modulator performs the oversampling and NS as it quantizes the amplitude of the input signal and filters the resulting quantization error. The quantizer bitstream v(n) is filtered to eliminate out-of-band noise and is down-sampled to the Nyquist-rate frequency [22].

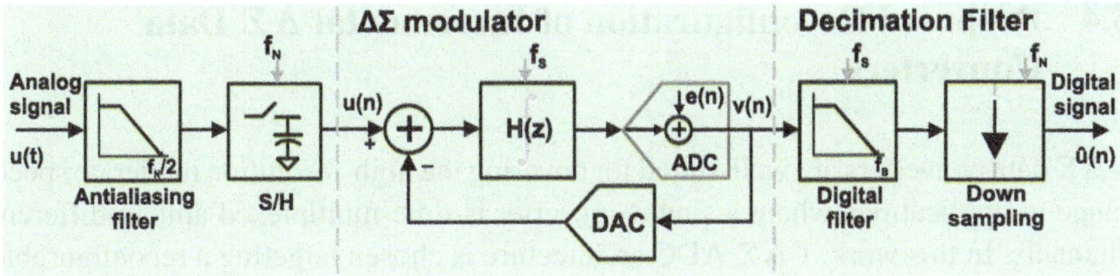

Fig. 5.6 Block diagram of a generic ΔΣ ADC adapted from [22]

In order to appreciate the oversampling and NS of ΔΣ modulators, a linearized quantizer model with additive quantization error e(n) is used to derive the signal (STF) and noise (NTF) transfer functions as follows [21, 22]:

$$Y(z) = STF(z) \cdot U(z) + NTF(z) \cdot E(z), \tag{5.2}$$
$$STF(z) = \frac{H(z)}{1 + H(z)}, \quad NTF(z) = \frac{1}{1 + H(z)}.$$

Ideally, the NTF completely removes the in-band noise without affecting the signal and its gain is allowed to increase outside the bandwidth of interest (i.e. high-pass filter characteristics) [22]. Conversely, the STF is desired to be as flat as possible within the bandwidth of interest and its gain drops outside the in-band region (i.e. low-pass filter characteristics). Applying this approach on (5.2) brings down the choice of H(z) to a Lth order integrator, which has large in-band gain and decreases outside the bandwidth of interest. A Lth order NTF is expressed as follows:

$$NTF(z) = \left(1 - z^{-1}\right)^{L}, \tag{5.3}$$

where the order of the NTF high-pass filter determines the amount of in-band noise suppression.

The theoretical Signal-to-Quantization-Noise Ratio (SQNR) for a third-order freely-running ΔΣ modulator with 1-bit quantizer is estimated as [23]:

$$SQNR\,(dB) = 10\log_{10}\left(\frac{21}{2\pi^{6}} OSR^{7}\right), \tag{5.4}$$

where OSR is the oversampling ratio $OSR = f_S/f_N$ and f_S being the modulator clock frequency. Note, that by doubling the OSR, up to 21 dB of SQNR improvements is achievable for a third-order freely-running ΔΣ modulator with 1-bit quantizer (compare to only 12 dB SQNR improvement for I-ΔΣ employing the same modulator). The aggressiveness of the NTF noise-shaping is upper-limited by the modulator's stability. As reported in [24], the stability of higher-order ΔΣ modulators is guaranteed as long as proper feed-forward coefficients are chosen and the out-of-band quantization noise gain is limited to 2 when an internal 1-bit quantizer is used.

5.4 Proposed Reconfiguration of Incremental $\Delta\Sigma$ Data Converters

I-$\Delta\Sigma$ data converters are well-suited for covering the high-resolution moderate speed range in applications where a single converter is time-multiplexed among different channels. In this work, I-$\Delta\Sigma$ ADC architecture is chosen targeting a reconfigurable resolution range from 12 to 16 b. For I-$\Delta\Sigma$ data converters, the $\Delta\Sigma$ modulator and digital filter are periodically reset and operate using an oversampling clock frequency f_S for a predetermined clock cycles M. Consequently, I-$\Delta\Sigma$ data converters are memoryless and nonlinear time-invariant systems that are among Nyquist-rate converters. Therefore, input signal frequencies f_{IN} that are within $f_S/(M \cdot 2) < f_{IN} < f_S/2$ alias back into the signal band [25, 26]. In the following section, the proposed reconfiguration architecture of I-$\Delta\Sigma$ ADCs is presented.

5.4.1 *Dynamic Range Presetting*

$\Delta\Sigma$ ADCs are intrinsically resolution and sample rate reconfigurable since the signal bandwidth is determined by the maximum sampling frequency f_S and the Oversampling Ratio (OSR, or M). However, a simple adjustment of the sampling frequency and/or the OSR without adapting the internal modulator structure would lead to largely varying energy efficiency versus resolution.

Figure 5.7 shows the block diagram of the proposed reconfigurable ADC operating at Nyquist conversion rate of $f_N = f_S/M$. The chosen architecture is a single loop, third-order cascaded-integrator-feed-forward (CIFF) $\Delta\Sigma$ modulator with a single-bit internal quantizer to achieve a good linearity at moderate OSRs. An off-chip chain of integrators (CoI) digital filter is used to extract the high-resolution digital signal from the quantizer bitstream. During a single conversion, the loop filter shapes the signal as well as the noise placing the most stringent requirements on the first integrator stage since its noise and distortion are not shaped by any preceding integrator stages.

In the presented design, the OSR of the I-$\Delta\Sigma$ modulator is reduced on system-level in the lower resolution modes thus, unlocking higher conversion rates. At the same time, the thermal-noise-limiting first integrator is scaled on circuit-level, allowing for a large reconfiguration in the ADC's dynamic range (DR). More specifically, *DR presetting* is achieved provided that a certain number of unit integrators S_{ON} is a priori selected from a maximum number of available unit integrators S_{MAX} and the associated modulator's OSR is optimally chosen according to the required DR. The selected unit integrators are then connected in parallel acting collectively as the modulator's first integrator throughout an entire conversion cycle. If an additional power saving is needed, the proposed reconfiguration approach can be combined with the *dynamic scaling of noise* method of [27] to further optimize the energy efficiency at higher resolution modes.

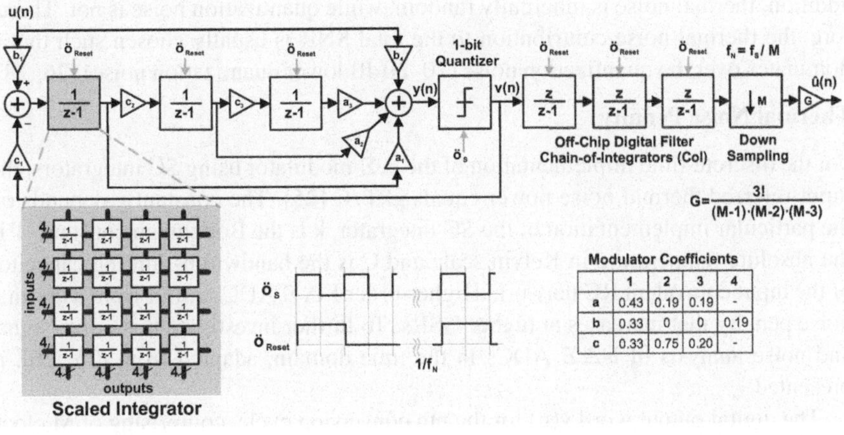

Fig. 5.7 Block diagram of the proposed reconfigurable I-ΔΣ ADC incorporating a 4-b scalable first integrator arranged in a switch matrix configuration

Quantization Noise

The accuracy of the proposed I-ΔΣ ADC is quantization noise limited for the lower resolution modes and thermal noise-limited for the higher resolution modes. In order to achieve power-efficient operation through all resolution modes, the noise behavior of I-ΔΣ ADCs is reviewed. Consequently, the optimal settings for the modulator's OSR and the number of active unit integrators are achieved.

In oversampling ΔΣ ADCs, the quantization noise is suppressed by the high-pass noise-shaping characteristics of the NTF. For a single loop third order I-ΔΣ ADC employing a single-bit quantizer and CoI digital filter, the theoretical SQNR is given by Caldwell and Johns[26]:

$$\mathrm{SQNR\ (dB)} = 6.02\log_2\left(\alpha\frac{(\mathrm{M}+2)!}{3!(\mathrm{M}-1)!}+1\right)+1.76, \tag{5.5}$$

where α is the maximum amplitude of the input signal that keeps the quantizer input bounded (cf. 5.4 for freely-running ΔΣ modulator). As an example, reducing the resolution from 14 b to 12 b, which is equivalent to reducing the SQNR by 12 dB, requires halving M that allows for doubling the conversion rate. In lower resolution modes, the lower bound for the input capacitor(s) of the first SC integrator is set and is usually chosen to ensure that the matching requirements of the modulators' coefficients are fulfilled. Consequently, the kT/C thermal noise requirement is easily met and the ADC operation is limited by the quantization noise.

For higher resolution modes, in order to lower the thermal noise, a higher-order ΔΣ modulator, higher OSR, or higher input capacitance value C are the available design decisions. Only the value of the input capacitor C does not determine the quantization noise, yet it primarily determines the ADC's power consumption. In addition, thermal noise is inherently random, while quantization noise is not. Therefore, the thermal noise contribution to the total SNR is usually chosen such that it dominates over the quantization noise (10–20 dB lower quantization noise) [26, 28].

Thermal Noise Penalty

For the discrete-time implementation of the ΔΣ modulator using SC integrators, the input-referred thermal noise power equals gkT/C [25]. The constant g depends on the particular implementation of the SC integrator, k is the Boltzmann constant, T is the absolute temperature in Kelvin scale and C is the bandwidth-limiting capacitor of the input equivalent RC network. Higher-order I-ΔΣ ADCs suffer from a thermal noise penalty that increases at higher OSRs. To further investigate this issue, signal and noise analysis of I-ΔΣ ADCs in the time domain, adapted from [25, 26], is presented.

The digital output word v(n) for the nth conversion cycle, comprising of M clock periods, can be calculated using the finite-length convolution operation as follows:

$$\begin{aligned} v(n) &= h(k) * d(k) \\ &\triangleq \sum_{k=nNM}^{[nN+1]M-1} h([nN+1]M-1-k) \cdot d(k), \end{aligned} \tag{5.6}$$

where h(k) is the impulse response of the digital filter and N is the number of channels. The digital output sequence of the quantizer d(k) is given by:

$$d(k) = stf(k) * [u(k) + t(k)] + ntf(k) * q(k), \tag{5.7}$$

where stf(k) and ntf(k) are the impulse response of the signal and noise transfer function of the ΔΣ modulator, u(k) is the input signal and t(k) and q(k) are the input-referred thermal and quantization noise, respectively. By combining (5.6) and (5.7), an end-to-end relation is given by:

$$v(n) = h(k) * stf(k) * [u(k) + t(k)] + h(k) * ntf(k) * q(k). \tag{5.8}$$

Using the finite-length convolution operation, the thermal noise power contribution to the output v(n) is estimated:

$$\begin{aligned} v_t(n) &= h(k) * stf(k) * t(k) = stf'(k) * t(k) \\ &\triangleq \sum_{k=nNM}^{[nN+1]M-1} stf'([nN+1]M-1-k) \cdot t(k), \end{aligned} \tag{5.9}$$

where stf′(k) is the impulse response of the end-to-end signal transfer function. The impulse response of stf′(k) can be described as $stf'[k] = \{s'_0, s'_1, \ldots, s'_{M-1}\}$. By choosing stf′(k) to equal 1 [26], the following constraint results:

$$\sum_{i=0}^{M-1} s_i' = 1. \tag{5.10}$$

The impulse response stf′(k) shapes the input-referred thermal noise, which results in the following thermal noise power spectral density:

$$\overline{v_t(n)^2} = \frac{gkT}{C} \sum_{i=0}^{M-1} s_i'^2. \tag{5.11}$$

By observing (5.11), the condition for minimum thermal noise contribution to the output v(n) demands $s_i' = 1/M$ for all values of i. This reduces (5.11) to equal $gkT/(M \cdot C)$, which is the same condition for the freely-running ΔΣ ADCs regardless of the modulator order. Conversely, this condition is normally not suitable for higher-order I-ΔΣ ADCs, since the last few output samples of d(k) contain large quantization errors and must be suppressed by a low-pass digital filter. Therefore, the impulse response of the digital filter h(k) is chosen to apply more weighting to earlier outputs d(k) than later ones, while maintaining the constraint given by (5.10) for a unity STF [26].

Considering a unity STF assumes a negligible effect for the ΔΣ modulator with a low-distortion implementation on the STF [25, 26, 29]. Therefore, the impulse response of the digital filter h(k) is the dominant shaping-response of the output v(n). Subsequently, the thermal noise relation (5.11) can be adjusted as follows:

$$\overline{v_t(n)^2} = \frac{gkT}{C} \sum_{k=0}^{M-1} h^2(k) = \beta \frac{gkT}{M \cdot C}. \tag{5.12}$$

A so-called thermal noise penalty β results from such uneven weighting of the samples' thermal noise by the digital filter and the deviation from the optimum even weighting of 1/M. The thermal noise penalty is calculated as [25]:

$$\beta = \frac{\text{actual thermal noise power}}{\text{minimum thermal noise power}} = M \sum_{k=0}^{M-1} h^2(k). \tag{5.13}$$

Fig. 5.8a plots the modeled OSR settings and thermal noise penalty of different resolution modes for the implemented third order I-ΔΣ ADC.

Figure 5.8b shows the number of unit integrators S_{ON} that are activated in order to achieve a certain DR for S_{MAX} from 1 to 32. The typical fixed-resolution I-ΔΣ ADC employs a whole first integrator ($S_{MAX} = 1$) and is always-on ($S_{ON} = 1$) regardless of the OSR value M. Alternatively, in the proposed *DR presetting* approach, the first integrator is scaled in order to match the thermal noise and power efficiency requirements of the high and low-resolution modes, respectively. Note, that the availability

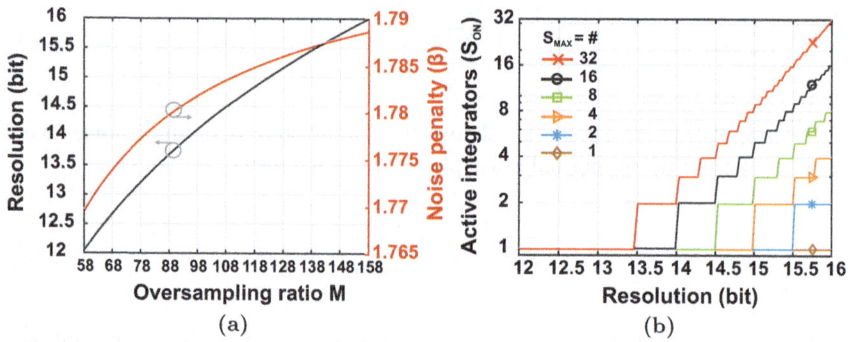

Fig. 5.8 ADC configuration using the DR presetting method for different resolution modes. **a** Oversampling ratio M and thermal noise penalty β for different resolution modes related to a modeled third order CIFF I-$\Delta\Sigma$ ADC utilizing CoI digital filter. **b** The number of active unit integrators S_{ON} versus the resolution range from 12 to 16 b for different S_{MAX}

of more hardware resources (i.e. higher S_{MAX}) provides the modulator with the reconfiguration flexibility to pinpoint the thermal noise suitable for the required DR.

Figures 5.9a, b show the modeling results of the relative change in FoM_S and FoM_W, respectively, for different resolution modes and S_{MAX} values. The FoMs are calculated based on the modeled behavior of third order I-$\Delta\Sigma$ ADCs and taking into consideration the lower OSR at lower resolutions shown in Fig. 5.8a. As an example, considering $S_{MAX} = 16$, if the DR is to be changed from 16 to 15 b, the *DR presetting* adjusts M from 158 to 123 (cf. Fig 5.8a) and S_{ON} from 16 to 5 (cf. Fig 5.8b). Therefore, the $\beta\, gkT/(C \cdot M)$ noise is increased by $10 \cdot \log((16 \times 158)/(5 \times 123)) = 6.1$ dB, which corresponds to a decrease in the ENOB by 1 b. It is interesting to observe that FoM_W is near constant for S_{MAX} higher than 8, which turns out to be sufficient number of unit integrators supporting a wide DR reconfiguration.

5.4.2 *Dynamic Scaling of Noise*

A straight-forward digital filter implementation is the Cascade (Chain) of Accumulators or Integrators (CoI) with the number of stages equivalent to the $\Delta\Sigma$ modulator order. Other implementations of the digital filter trades off higher OSRs for attractive features, such as suppression of periodic noise [30]. The terms h_i describe the impulse response of the digital filter, which is reset after M periods, as

$$h(k) = \{h_0, h_1, ..., h_{M-1}\}.$$

The terms h_i for the first- and second-order CoI digital filters simply equal $h_i = 1$ and $h_i = (i + 1)$, respectively. For a third-order CoI digital filter, the terms h_i are given by:

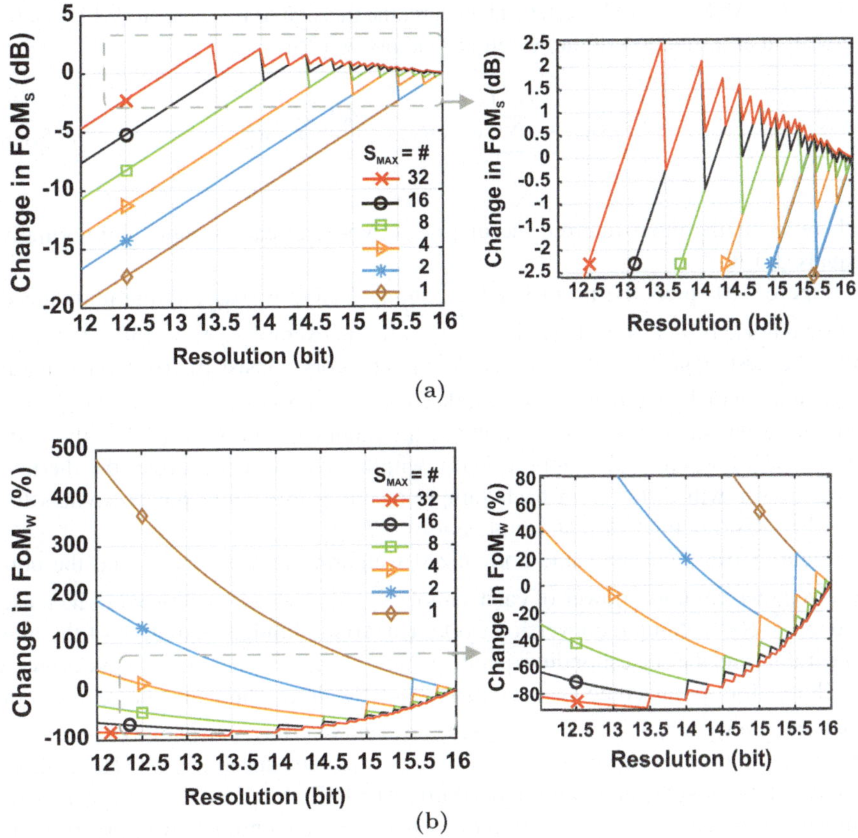

Fig. 5.9 Modeling results of the relative change in **a** FoM_S and **b** FoM_W using the proposed DR presetting method

$$h_i = \sum_{i=0}^{k} i + 1. \tag{5.14}$$

From (5.6), the quantizer output d(k) is convoluted with the impulse response h(k) and for higher-order filters, the earlier d(k) are more weighted than later samples [26]. To illustrate this uneven weighting, weighting factors w_i are derived from the filter's impulse response:

$$\begin{aligned} w_i &= \{w_0, w_1, ..., w_{M-1}\} \\ &= \frac{1}{g}\{h_{M-1}, h_{M-2}, ..., h_0\}, \end{aligned} \tag{5.15}$$

where $g = M(M+1)(M+2)/6$. Therefore, the thermal noise relation (5.12) can be re-written as a function of the weighting factors w_i [26]:

$$\overline{v_t(n)^2} = \sum_{i=0}^{M-1} w_i^2 \cdot \overline{v_{s,i}^2} = \frac{gkT}{C} \sum_{i=0}^{M-1} w_i^2, \tag{5.16}$$

where $\overline{v_{s,i}^2}$ is the input-referred thermal noise power of each sample and normally equals gkT/C.

Figure 5.10a plots the normalized weighting factors w_i^2/w_0^2 and the normalized thermal noise power of each sample $\overline{v_{s,i}^2}/\overline{v_{s,0}^2}$ for an exemplary $M = 158$. Traditionally, the first-stage SC integrator of $\Delta\Sigma$ modulators consist of fixed-value input capacitors, which results in a constant thermal noise power of each sample $\overline{v_{s,i}^2}$ (i.e. time invariant sample noise). However, if the input capacitors, as well as the first-stage integrator, are scaled appropriately within the conversion cycle, the thermal noise power will differ from each sample resulting in a more power-savings for I-$\Delta\Sigma$ ADCs (i.e. time variant sample noise) [27, 29].

Figure 5.10b plots the same normalized weighting factors w_i^2/w_0^2 and the normalized thermal noise power of each sample $\overline{v_{s,i}^2}/\overline{v_{s,0}^2}$, which is allowed to increase progressively within the conversion cycle. Due to the limited resolution of the unit integrators and the corresponding quantized thermal noise, switching instances are to be chosen for optimized power savings versus SNR drop. The scaled noise power of each sample $\overline{v_{s,i}^2}/\overline{v_{s,0}^2}$ is now a stair-case shaped curve. The concept of dynamic scaling of thermal noise within a conversion cycle was introduced in [27]. The authors used a 2-bit scaled SC integrator into a third-order CIFF I-$\Delta\Sigma$ ADC. In [29], an optimization problem was formulated in order to choose the optimal switching instants so that FoM_S is maximized.

In this work, an intuitive and generic approach is proposed by observing the fact that the thermal noise power of each sample is mainly shaped by the impulse response of the digital filter h(k) and when allowed to increase in a similar fashion to the drop in the digital filter weighting factors w_i^2, the optimal switching instants (i) can directly be extracted. To accomplish this, the order of the weighting factors w_i is reversed resulting in new weighting factors m_i given by:

$$\begin{aligned} m_i &= \{m_0, m_1, ..., m_{M-1}\} \\ &= \{w_{M-1}, w_{M-2}, ..., w_0\}. \end{aligned}$$

These new weighting factors are then applied to the thermal noise power of each sample so that they change in a way that is inherently compatible with the impulse response of the digital filter :

$$\overline{v_t(n)^2} = \sum_{i=0}^{M-1} w_i^2 \cdot \overline{v_{s,i}^2} = \frac{gkT}{C} \sum_{i=0}^{M-1} w_i^2 \cdot (1 + r \cdot m_i^2), \tag{5.17}$$

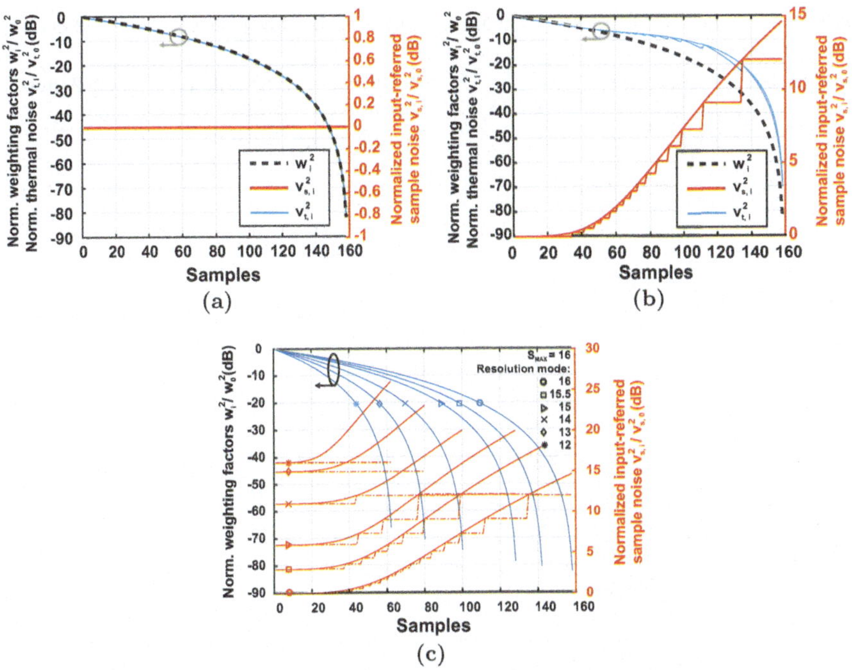

Fig. 5.10 Dynamic scaling of noise concept illustrated by plotting the normalized weighting factors w_i^2/w_0^2 and normalized thermal noise of each sample. **a** The typical case wherein each sample contributes a constant level of thermal noise. **b** Dynamic scaling of the thermal noise of each sample as it is allowed to increase progressively in a way that is inherently compatible to the impulse response of the digital filter. **c** Combining DR presetting method with the dynamic noise scaling technique. Normalized weighting factors of the digital filter w_i^2/w_0^2 (left y-axis) and the normalized thermal noise power of each sample (right y-axis) are plotted using optimal switching instances for multiple resolution modes namely, 16, 15.5, 15, 14, 13 and 12 b

where the typical condition of a constant thermal noise power of each sample occurs when the scaling parameter $r = 0$. The first row of Table 5.4 lists the parameters used for generating the example shown in Fig. 5.10b and the expected decrease in SNR and power consumption.

Figure 5.11a, b plot the decrease in the power consumption versus the drop in SNR (i.e. increase in total noise power) for different numbers of available integrators units (S_{MAX}). The corresponding FoM_S and FoM_W are also plotted on the right y-axes. Note, as the number of available integrator units increase, the step size of the sample noise power decreases thus, the rise in thermal noise is able to better approximate and follow the fall in the digital filter weighting factors. Additionally, an improvement of up to 2.3 dB and −48% can be ideally expected to the FoM_S and FoM_W from a scaled integrator with $S_{MAX} = 16$ units, respectively. For a scaled integrator with $S_{MAX} = 4$ units, an improvement of up to 1.9 dB and −40% can be ideally expected to the FoM_S and FoM_W, respectively. This is in agreement with the results reported in [29].

Table 5.4 Modulator settings and optimal switching instances when DR presetting and dynamic scaling of noise are used

ENOB (bits)	$S_{ON}(unit)$	M	Switching instance i	SNR drop (dB)	SQNR drop (dB)
16	$S_{MAX} = 16$	158	35, 42, 48, 52, 57, 61 65, 69, 73, 78, 84 91, 99, 112, 135, x	-0.8	0
15.5	9	142	20, 38, 47, 54, 61 69, 80, 99, x	-3.7	-3
15	5	128	19, 45, 58, 77, x	-6.4	-6
14	2	100	44, x	-11	-12
13*	1	80	x	-15	-18
12*	1	62	x	-16.1	-24

x No unit integrator is disconnected
* Quantization-noise-limited mode

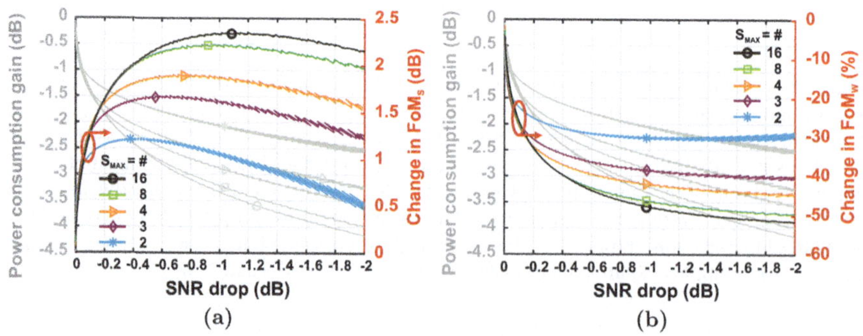

Fig. 5.11 The power-savings achieved using the dynamic scaling of noise plotted versus the increase in SNR. Changes in **a** FoM_S and **b** FoM_W are plotted for different values of S_{MAX}

Combined DR Presetting and Dynamic Noise Scaling

For the proposed 16–12 b resolution reconfigurable I-$\Delta\Sigma$ ADC, the *dynamic scaling of noise* is optionally combined with the *DR presetting* technique. Figure 5.10c plots the normalized weighting factors of the digital filter w_i^2/w_0^2 and the normalized thermal noise power of each sample for the resolution modes of 16, 15.5, 15, 14, 13 and 12 b. Note, that the *dynamic scaling of noise* is beneficial in the higher resolution modes since the ADC operation is thermal noise limited. Table 5.4 lists the optimal switching instances used for the different resolution modes shown in Fig. 5.10c

References

1. Murmann B (2019) ADC performance survey 1997–2019 (Online). Available: http://www.stanford.edu/~murmann/adcsurvey.html
2. Yip M, Chandrakasan AP (2011) A resolution-reconfigurable 5-to-10b 0.4-to-1V power scalable SAR ADC. In: 2011 IEEE international solid-state circuits conference, Feb 2011, pp 190–192
3. Harpe P, Zhang Y, Dolmans G, Philips K, De Groot H (2012) A 7-to-10b 0-to-4MS/s flexible SAR ADC with 6.5-to-16fJ/conversion-step. In: 2012 IEEE international solid-state circuits conference, Feb 2012, pp 472–474
4. El-Halwagy W, Mousavi P, Hossain M (2018) A 100-MS/s–5-GS/s, 13-5-bit nyquist-rate reconfigurable time-domain ADC. IEEE Trans Very Large Scale Integr (VLSI) Syst 26(10):1967–1979
5. Marien H, Steyaert MSJ, van Veenendaal E, Heremans P (2011) A fully integrated $\Delta\Sigma$ adc in organic thin-film transistor technology on flexible plastic foil. IEEE J Solid-State Circ 46(1):276–284
6. Schreier R (2019) Delta Sigma toolbox, MATLAB Central File Exchange (Online). Available: https://www.mathworks.com/matlabcentral/fileexchange/19-delta-sigma-toolbox
7. Analog-digital conversion. Analog Devices, Inc. (2004)
8. Nyquist H (1924) Certain factors affecting telegraph speed. Bell Syst Tech J 3:324–4346
9. Shannon CE (1948) A mathematical theory of communication. Bell Syst Tech J 27:623–656
10. Reeves AH (1942) "Electric signaling system. U.S. Patent 2,272,070
11. Goodall WM (1947) Telephony by pulse code modulation. Bell Syst Tech J 26:395–409
12. Doris K (2017) Hybrid data converter. In: 2017 workshop on advances in analog circuit design (AACD), March 2017, pp 1–15
13. Shu Y, Kuo L, Lo T (2016) An oversampling SAR ADC with DAC mismatch error shaping achieving 105 dB SFDR and 101 dB SNDR over 1 kHz BW in 55 nm CMOS. IEEE J Solid-State Circ 51(12):2928–2940
14. Yousry R, Chen M, Chang MF, Yang CK (2013) An architecture-reconfigurable 3b-to-7b 4GS/s-to-1.5GS/s ADC using subtractor interleaving. In: 2013 IEEE Asian solid-state circuits conference (A-SSCC), Nov 2013, pp 285–288
15. Fateh S, Schönle P, Bettini L, Rovere G, Benini L, Huang Q (2015) A reconfigurable 5-to-14 bit SAR ADC for battery-powered medical instrumentation. IEEE Trans Circuits Syst I Regul Pap 62(11):2685–2694 Nov
16. Rhee C, Kim S (2019) 8.4-to-16-bit resolution, 1-to-16 kHz bandwidth ADC with programmable-gain functionality for multi-sensor applications. Electron Lett 55(18):982–984
17. Chandrashekar K, Corsi M, Fattaruso J, Bakkaloglu B (2010) A 20-MS/s to 40-MS/s reconfigurable pipeline ADC implemented with parallel OTA scaling. IEEE Trans Circ Syst II Expr Briefs 57(8):602–606
18. (2020) American Semiconductor Homepage (Online). Available: https://www.americansemi.com/flex-ics.html
19. Papadopoulos NP, De Roose F, van der Steen JPJ, Smits ECP, Ameys M, Dehaene W, Genoe J, Myny K (2018) Toward temperature tracking with unipolar metal-oxide thin-film SAR C-2C ADC on plastic. IEEE J Solid-State Circ 53(8):2263–2272 Aug
20. Garripoli C, van der Steen JPJ, Smits E, Gelinck GH, Van Roermund AHM, Cantatore E (2017) 15.3 an a-IGZO asynchronous delta-sigma modulator on foil achieving up to 43dB SNR and 40dB SNDR in 300Hz bandwidth. In: 2017 IEEE international solid-state circuits conference (ISSCC), Feb 2017, pp 260–261
21. Inose H, Yasuda Y, Murakami J (1962) A telemetering system by code modulation - $\Delta\Sigma$ modulation. IRE Trans Space Electron Telemetry SET-8(3):204–209
22. Ortmanns M, Gerfers F (2006) Continuous-time Sigma-delta A/D conversion. Springer, Berlin
23. Tao S (2015) Power-efficient continuous-time incremental sigma-delta analog-to-digital converters. Ph.D. dissertation, KTH Royal Institute of Technology

24. Chao KC, Nadeem S, Lee WL, Sodini CG (1990) A higher order topology for interpolative modulators for oversampling A/D converters. IEEE Trans Circ Syst 37(3):309–318
25. Steensgaard J, Zhang Z, Yu W, Sarhegyi A, Lucchese L, Kim D, Temes GC (2008) Noise-power optimization of incremental data converters. IEEE Trans Circ Syst I Regul Pap 55(5):1289–1296
26. Caldwell TC, Johns DA (2010) Incremental data converters at low oversampling ratios. IEEE Trans Circ Syst I Regul Pap 57(7):1525–1537
27. Vogelmann P, Haas M, Ortmanns M (2018) A 1.1 mW 200 kS/s incremental $\Delta\Sigma$ adc with a DR of 91.5dB using integrator slicing for dynamic power reduction. In: 2018 IEEE international solid-state circuits conference—(ISSCC), Feb 2018, pp 236–238
28. Understanding Delta-Sigma Data Converters, 1st edn. Wiley-IEEE Press (2005)
29. Vogelmann P, Wagner J, Haas M, Ortmanns M (2019) A dynamic power reduction technique for incremental $\Delta\Sigma$ modulators. IEEE J Solid-State Circ 54(5):1455–1467 May
30. Markus J, Silva J, Temes GC (2004) Theory and applications of incremental $\Delta\Sigma$ converters. IEEE Trans Circ Syst I Regul Pap 51(4):678–690

Chapter 6
Reconfigurable Analog-to-Digital Converter for HySiF: Part II

In this chapter, the circuit-level implementation of the proposed reconfigurable I-$\Delta\Sigma$ ADC is reported. The chapter starts with discussing few practical system-level design considerations, such as dividing the first SC integrator into scalable unit elements and the impact of mechanical stress on the ADC performance. Next, transistor-level implementation of critical building blocks are adapted to serve the power efficiency and reconfigurability of the ADC. Furthermore, considerations for the floor-planning and physical design of the ADC chip are discussed taking into account the distributed and scalable nature of the first SC integrator. Finally, measurement results are presented for the bulk and ultra-thin ADC chips.

6.1 System Design Considerations

Fully-Differential (FD) configuration is employed in this work due to its several advantages compared to the single-ended one. Systems that utilize FD circuitry benefit from improved common mode and power supply noise rejection and reduced charge injection effects [1].

Figure 6.1 shows the implemented reconfigurable I-$\Delta\Sigma$ ADC and the associated timing diagram. In order to render the entire first integrator stage scalable with resolution, the sampling and integration capacitors, as well as the switching network and the OTA are designed as unit elements. Furthermore, the load capacitance seen by the ADC driver also scales with resolution. The structures of the SC integrators in the second and third stages resemble those of the first stage unit integrators, with the only difference being the corresponding capacitor ratios. Consequently, the design of the $\Delta\Sigma$ modulator is reduced to the design of one unit SC integrator which is repetitively utilized as a standard cell (i.e. 16$\times$ cells for the first integrator, 1$\times$ cell for the second, as well as the third integrator).

M. Elsobky, *Ultra-Thin Sensors and Data Conversion Techniques for Hybrid System-in-Foil*, Recognizing Outstanding Ph.D. Research,
https://doi.org/10.1007/978-3-030-97726-9_6

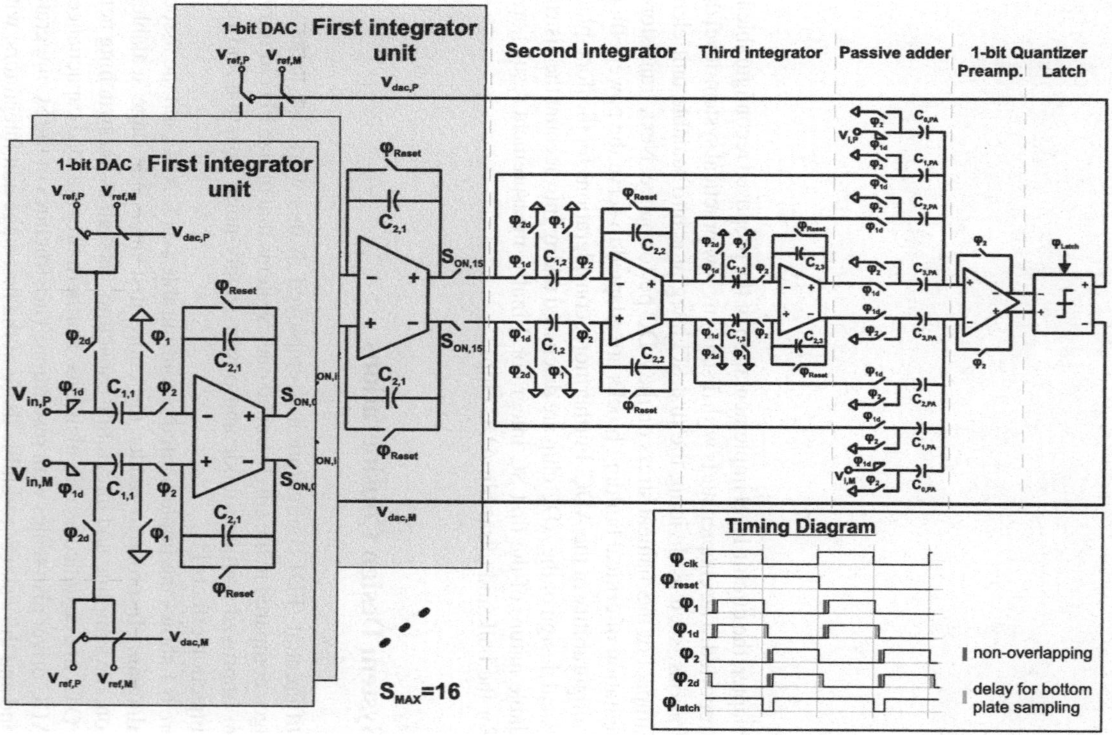

Fig. 6.1 Schematic of the implemented reconfigurable IΔΣ ADC. The input switches' bootstrap circuit and the CMFB circuit are shared between the unit integrators

6.1.1 Switched Capacitor Integrator

In Fig. 6.1, a FD parasitic-insensitive SC integrator is used in the first, second and third stages of the proposed I-ΔΣ ADC. This basic SC integrator is operated using a clocking scheme with two phases, namely Φ_1 and Φ_2, in addition to a reset phase Φ_{reset}. In I-ΔΣ modulators, the reset phase is triggered at the Nyquist conversion rate $f_N = M/T_S$. The operation of the SC integrator ($\Phi_{reset} = 0$) is described using the following transfer function:

$$H(z) = \frac{C_1/C_2 z^{-1}}{1 + 1/A(1 + C_1/C_2) - (1 + 1/A)z^{-1}} \approx \frac{C_1 z^{-1}}{C_2(1 - z^{-1})}, \tag{6.1}$$

where C_1 and C_2 correspond to the sampling and integration capacitors, respectively.

During Φ_{reset}, the voltage on C_2 is ideally zero thus, the memory of the integrator is cleared. During Φ_1, the input V_{in} is sampled on the bottom plate of the capacitor C_1, while the integration capacitor C_2 holds on the charges from the previous clock cycle. During Φ_2, the input is disconnected and the reference V_{ref} for the first stage or the common mode voltage V_{cm} for the second and third stages, are connected as the integrator accumulates the charges from C_1 to C_2 with a voltage gain of C_1/C_2.

Here, one drawback is due to the charge transfer between the input and reference, which occurs if the input and reference have different common mode voltages [2]. On the other side, an advantage is the validity of the output voltage V_{out} at the end of both phases, which provides a flexible implementation for the clocking scheme in order to ensure power-efficient operation. For instance, the output voltage changes only during Φ_2. Therefore, the capacitive load of the OTA during phase Φ_2 is minimized as it sets the overall dynamic characteristics, such as SR and Gain-Bandwidth (GBW). Additionally, the subsequent stage samples the output voltage and loads the OTA during the less critical phase Φ_1.

6.1.2 Capacitor Sizing

Targeting a maximum SNR of 100 dB, the mean-square noise voltage relative to $\alpha = 0.68$ (i.e. the maximum amplitude of the input signal that keeps the quantizer input bounded) is calculated using the following relation [3]:

$$\overline{v_n^2} = \frac{(\alpha V_{DD}/2)^2/2}{10^{SNR/10}} = \frac{(0.68 \times 1.65)^2}{2 \times 10^{10}} = (7.9 \times 10^{-6})^2. \tag{6.2}$$

The relation in (5.12) sets the kT/C thermal noise of DT I-ΔΣ ADCs. Here, the thermal noise contribution of the second and third integrators to the total SNR is neglected since their noise is shaped by the previous stage(s). Therefore, the total sampling capacitor $C_{1,total}$ of the first integrator is calculated from the in-band input referred mean-square noise voltage using the following relation:

$$C_{1,\text{total}} = S_{MAX} \cdot C_{1,\text{unit}} = \beta \frac{gkT}{M \cdot \overline{v_n^2}} = 2 \frac{2kT}{158 \times (7.9 \times 10^{-6})^2} = 1.67\,\text{pF}. \quad (6.3)$$

Assuming equal noise contribution from each unit integrator, the lower limit for the sampling capacitor $C_{1,\text{unit}}$ of each unit integrator is the kT/C noise requirement for the highest resolution mode. Another lower limit is determined by the mismatch requirements in order to keep the integrator coefficients of all unit integrators sufficiently close. Therefore, the unit sampling capacitor $C_{1,\text{unit}}$ is chosen to be $2\times$ 106 fF for S_{MAX} of 16 unit integrators. The same unit capacitors are used to realize the coefficients of the SC integrators for the second and third stages.

6.1.3 Large-Signal Settling

At the start of Φ_2, the non-linear settling starts as the SC integrator charges the integration capacitor using the maximum current supported by the OTA's output stage. As a rule of thumb for power-efficient slew rate limited design, the duration of the non-linear settling is allocated about one half of Φ_2 (i.e. $T_S/4$) [3]. By using the clocking scheme shown in Fig. 6.1, the OTA of the first integrator is not loaded by the second stage or the passive adder during the critical clock phase Φ_2. However, in this phase, the OTA is loaded by its sampling and integration capacitors, as well as the capacitors of the SC Common Mode Feedback (CMFB) circuit C_{CMFB} and circuit interconnect parasitic capacitors C_P. The following relation summarizes the overall output-referred load capacitance $C_{\text{load,total}}$ during Φ_2

$$C_{\text{load,total}} = S_{MAX} \left(\frac{C_{1,\text{unit}} C_{2,\text{unit}}}{C_{1,\text{unit}} + C_{2,\text{unit}}} \right) + C_{CMFB} + C_P. \quad (6.4)$$

Consequently, the maximum current $I_{\text{bias,max}}$ supported by the OTA's output stage is designed based on the following relation:

$$I_{\text{bias,max}} = C_{\text{load,total}} \times SR_{\text{total}} = C_{\text{load,total}} \times \frac{\max(V_{\text{out,swing}})}{T_S/4}, \quad (6.5)$$

where SR corresponds to the slew rate of the OTA and $\max(V_{\text{out,swing}})$ is the maximum amplitude swing for the first integrator output voltage.

Since non-linear settling is dependent on the instantaneous state of the $\Delta\Sigma$ modulator, statistical analysis of the integrators' output swings is needed. Figure 6.2a–c and d–f plot the probability of a certain amplitude and amplitude swings, respectively, for the first, second and third integrator. Here, the I-$\Delta\Sigma$ modulator is modeled using 1-bit internal quantizer whose decision is bound to ± 1. Therefore, the maximum amplitude swing $\max(V_{\text{out,swing}})$ for the first integrator, which uses a supply

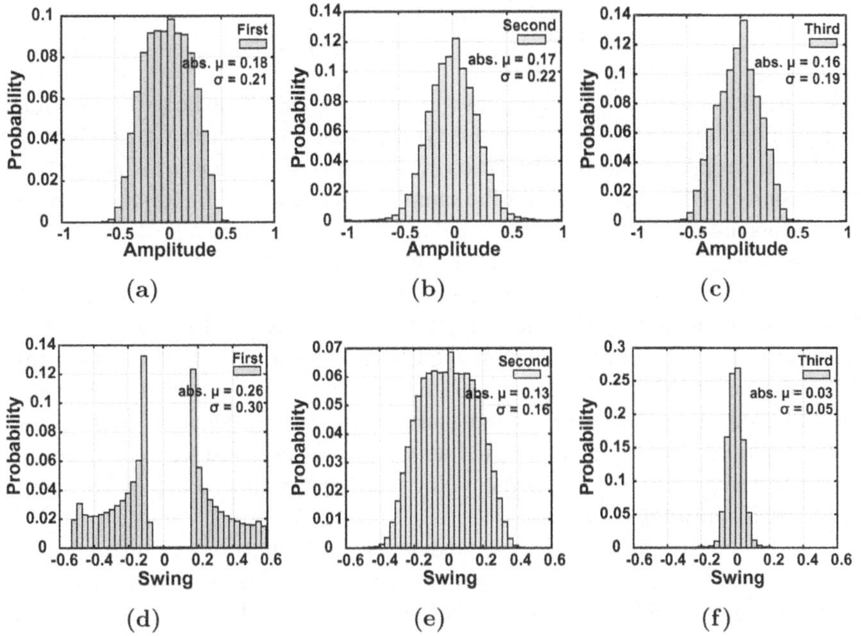

Fig. 6.2 Statistical analysis of the **a–c** amplitude and **d–f** amplitude swings of the first, second and third integrators. Here, a third order CIFF I-$\Delta\Sigma$ with 1-bit internal quantizer (decision is bound to ± 1) is modeled

voltage of 3.3 V is about 2V. Using (6.5), the maximum current supported by the OTA's output stage within each unit integrator equals about 6 μA. Here, equal driving capabilities for the 16 unit integrators is assumed.

Note, the output amplitudes of the first, second and third integrators are similarly limited to about ± 0.6 with close absolute mean (abs. μ) and standard deviation σ. Alternatively, the maximum output amplitude swing decreases from about ± 0.6 to ± 0.2 from the first to the third integrator, respectively. This directs at the reduced slew rate requirements for the OTAs incorporated in the second and third integrators.

6.1.4 Small-Signal Settling

After the non-linear settling, the OTA takes control and determinately completes the charge-transfer in an exponential fashion. The OTA can be modeled as a first order low pass filter within its GBW with a RC time constant τ. The total transconductance $g_{m,total}$ of the first integrator's OTA is calculated using the following relation:

$$g_{m,total} = \frac{C_{load}}{\tau} = \frac{C_{load}}{T_S/4 \cdot 1/n}, \tag{6.6}$$

where n is the number of time constants τ required to reach adequate settling accuracy. For high resolution DT $\Delta\Sigma$ ADCs, n is typically chosen between 12 and 20. In this work and supported by the simulation results, n = 16 is chosen, which sets the OTA's unity GBW to be ten times the sampling frequency ($10 \times f_S$). Since the unit integrators are connected in parallel, the active unit $g_{m,unit}$ are added results in the maximum $g_{m,total} = S_{MAX} \cdot g_{m,unit}$ for the high resolution mode.

Due to the fact that less unit integrators of the first stage are activated towards the lower resolution modes, the ratio of the CMFB capacitors to the total load capacitance increases. In a similar situation, the CMFB capacitors of the second and third integrators are not shared and present a rather large ratio of the total load capacitance. Here, the reduction of the OTAs' driving and settling capabilities are not critical since those requirements are already relaxed for the second and third stages. For the case of the first integrator, the requirements for speed and settling accuracy are relaxed only towards lower resolutions.

These deterministic variations of the dynamic performance of the first integrator are simulated using transient and stability analysis in Cadence Virtuoso with the proposed circuit configuration shown in Fig. 6.3. The stability analysis is triggered at the end of the clock phase Φ_2. The simulated transient response of S_{ON} active integrators, as well as the AC response of the associate OTAs are plotted in Fig. 6.4. For instance, during the highest resolution mode ($S_{ON} = 16$), the ratio of the CMFB capacitor to the total load capacitance is negligible ($\approx 1/17$). Therefore, the collective performance of the active OTAs result in the highest GBW and SR values.

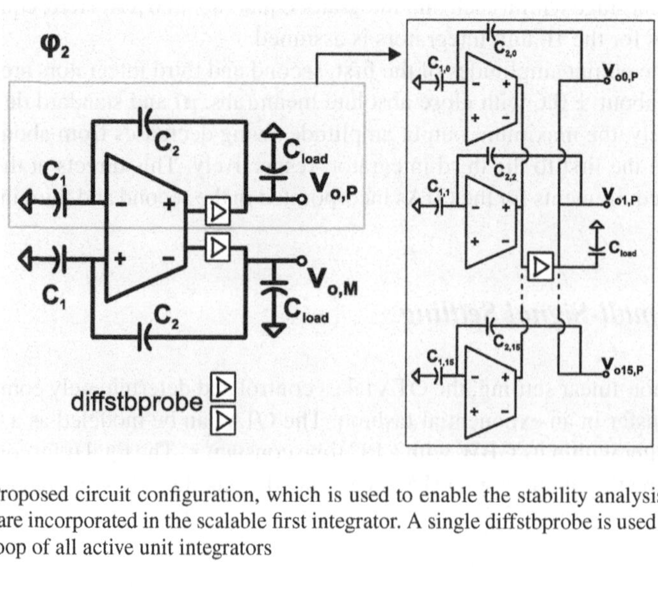

Fig. 6.3 Proposed circuit configuration, which is used to enable the stability analysis of the unit OTAs that are incorporated in the scalable first integrator. A single diffstbprobe is used to break the feedback loop of all active unit integrators

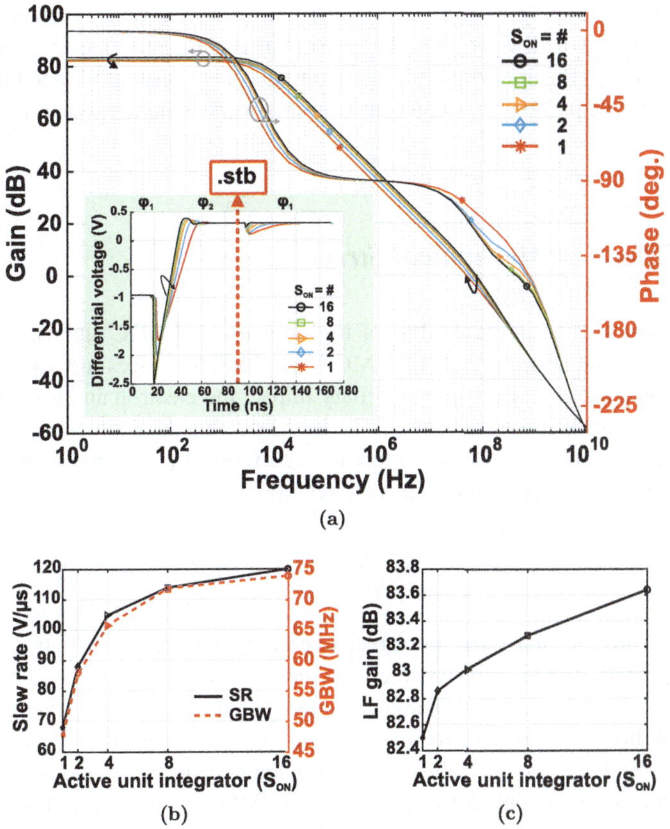

Fig. 6.4 **a** Simulated AC response for different number of active unit integrators S_{ON}. The stability analysis (.stb) is performed using the arrangement shown in Fig. 6.3. The inset shows the transient response for the same number of unit integrators. Here, the stability analysis is triggered at the end of Φ_2. **b** The extracted slew rate, GBW and **c** low frequency (LF) gain of the OTA's are plotted versus S_{ON}

However, during the low resolution modes when only one unit integrator is active ($S_{ON} = 1$), this capacitance ratio increases to about 1/2. Consequently, the OTA's GBW and SR decrease. It is important to note, that the OTA's DC gain is maintained above 80 dB regardless of the number of active integrators. This is due to the fact that the product of g_m and the OTA's output resistance r_{ds} is near-constant according to the following relation:

$$A_v = g_{m,total} \cdot r_{ds,total} = S_{ON} g_{m,unit} \frac{r_{ds,unit}}{S_{ON}} = g_{m,unit} \cdot r_{ds,unit}. \tag{6.7}$$

It is interesting to point out that a similar approach has recently been investigated to dynamically reduce the OTA's SR and GBW towards the end of a single A/D conversion in an attempt to achieve power-savings [4]. However, such reduction in SR and GBW drastically degraded the performance of the presented high-resolution ADC. This proves the importance of maintaining a near-constant SR and GBW when the proposed DR presetting is combined, in this work, with the dynamic scaling of noise.

6.1.5 Effects of Mechanical Stress

In order to gauge the susceptibility of the implemented ADC against mechanical stress σ, the simulation models of NMOS and PMOS devices are varied. In fact, the variations of mechanical stress primary impact the electron and hole mobilities, thereby changing the drain currents [5–7]. By neglecting the geometrical and threshold voltage variations [5], the change in drain current ΔI_D is related to the change in the carrier mobility $\Delta\mu$ by the following relation:

$$\frac{\Delta I_D}{I_D} \approx \frac{\Delta\mu}{\mu} = -\Pi_{ijkl} \cdot \sigma_{kl}, \tag{6.8}$$

where Π_{ijkl} are the first-order piezoresistive coefficients and σ_{kl} is the stress tensor [5, 7]. Since the piezoresistive coefficients of the utilized CMOS technology are not available, the mobility changes due to in-plane stress in the longitudinal and transversal directions are selected based on previous work on silicon UTCs (0.5 μm CMOS) [5–8]. Consequently, the device parameters *ne3i_u0_m* and *pe3i_u0_m* are varied in the BSIM models for the NMOS and PMOS, respectively. Table 6.1 summarizes the effects of applying an exemplary uniaxial stress magnitude of ±400 MPa on the mobility change in NMOS and PMOS, when the devices are oriented in the transversal and longitudinal with respect to the same crystalline orientation [110] of a conventional (001) silicon wafer. The +/− signs represent the tensile and compressive stress, respectively.

Figure 6.5a, b plot the simulated SQNR of the ADC ($S_{MAX} = 16$) versus variations in the applied uniaxial longitudinal and transverse stress, respectively. For instance, if the on-resistance of TGs is observed, one can detect the drop in resistance due to the application of tensile or compressive longitudinal stress as either the mobility of PMOS or NMOS decreases. Consequently, the SQNR significantly drops for stress values higher than 1.6 GPa and 2.8 GPa for tensile and compressive longitudinal stress, respectively. When the applied stress is transversal to the channel, the change in mobility is similar in NMOS and PMOS devices (similar trend, yet different magnitudes). Therefore, the SQNR does not drop while increasing the carrier mobility but it drops for stress values higher than 2 GPa for compressive transversal stress.

Table 6.1 Mobility change in NMOS and PMOS for an exemplary stress magnitude of ±400 MPa

Condition	NMOS	PMOS	Condition	NMOS	PMOS
stress magnitude = +400 MPa stress direction = [110] channel orientation = [110]	+12%	−17%	stress magnitude = +400 MPa stress direction = [110] channel orientation = [$\bar{1}$10]	+6%	+16%
stress magnitude = -400 MPa stress direction = [110] channel orientation = [110]	−12%	+17%	stress magnitude = −400 MPa stress direction = [110] channel orientation = [$\bar{1}$10]	−6%	−16%

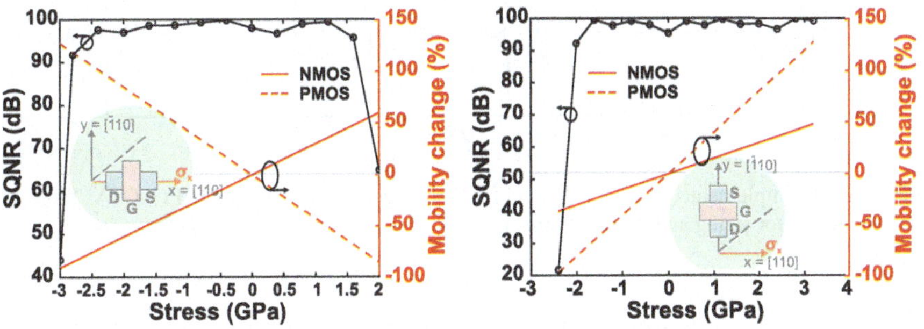

(a) (b)

Fig. 6.5 Effect of mechanical stress on the accuracy of the implemented ADC. The SQNR is plotted when NMOS and PMOS devices are oriented along the **a** longitudinal and **b** transversal direction to the applied in-plane stress using the same crystalline orientation [110] of a conventional (001) silicon wafer. The +/− signs represent the tensile and compressive stress, respectively

6.2 Circuit Design

6.2.1 *Operational Transconductance Amplifier*

Each unit integrator contains a FD OTA, which is implemented as a FD version of the single-ended recycling folded cascode (RFC) OTA reported in [9]. Figure 6.6 shows the transistor-level design of the implemented FD RFC OTA. The input pair M1 and M2 are DC biased using constant current source M0 ($2I_{bias}$). Furthermore, they are operated in the moderate inversion region in order to strike a balance between achieving high power efficiency $g_m/I_D \approx 15$ and a moderate gate-drain C_{gd} capacitance (as C_{gd} additionally loads the OTA).

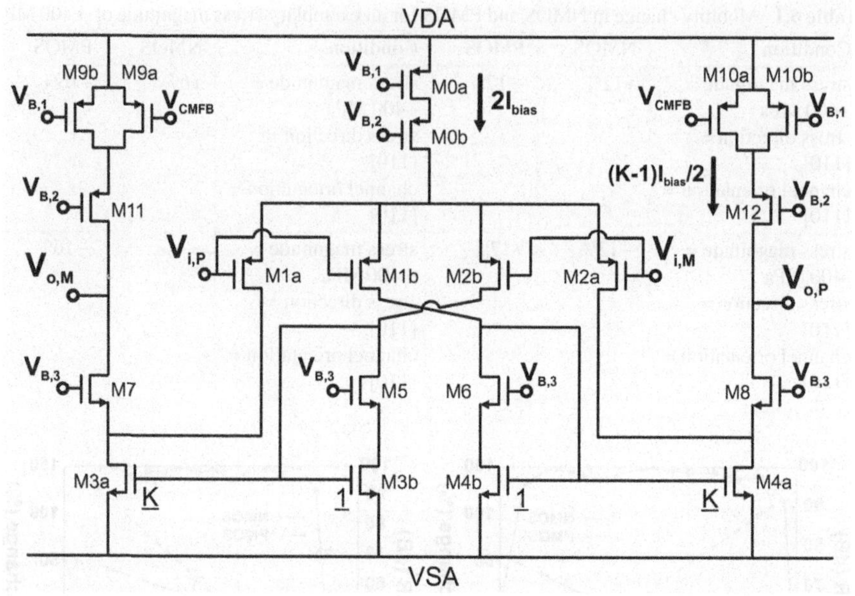

Fig. 6.6 The implemented fully differential (FD) recycling folded cascode (RFC) OTA. Table 6.2 lists the transistor sizing and drain currents I_D

Compared to the conventional folded cascode OTA, the input pair are divided into two halves, namely M1a, M2a, M1b and M2b. The latter two are cross-coupled and connect to recycling current paths, which are formed by wide-swing current mirrors (M3–M8) with current mirroring gain K. Hence, M3 and M4 are not only used as DC current sources but also as AC drivers for the output branches. Finally, M9-M12 are cascode current sources in which half of M9 and M10 are used for the control of the common mode output level, thus ensuring the operational symmetry of the FD OTA. Table 6.2 lists the sizes and drain currents I_D of the transistors, which are used to implement one unit element of the FD RFC OTA.

The large-signal and small-signal analysis of the single-ended RFC [9] also apply for the implemented FD RFC OTA. The SR and g_m are enhanced by a factor of K and (1+K), respectively, which are given by the following relations:

$$SR_{unit} = \frac{K \cdot 2I_{bias,unit}}{C_{load,unit}}, \tag{6.9}$$

$$g_{m,unit} = g_{m,1} = g_{m,1a}(1 + K). \tag{6.10}$$

In the recycling path, a higher current mirroring gain K improves g_m, GBW and SR but degrades the phase margin and input referred noise. In practice, less K value is achievable due to the discrepancy in the current mirroring during large- and

Table 6.2 Transistors sizing of the implemented FD RFC OTA

Transistor	I_D (μA)	W/L (μm/μm)
M0a	12	28.16/1.8
M0b	12	56.32/0.45
M1a,M1b, M2a,M2b	4	14/0.3
M3a,M4a	6	1.76/1.8
M3b,M4b	3	0.88/1.8
M5,M6	3	3.52/0.45
M7,M8	3	3.52/0.45
M9a,M10a	2.25	5.28/1.8
M9b,M10b	0.75	1.76/1.8
M11,M12	3	14.08/0.45

Table 6.3 Performance summary of the implemented FD RFC OTA

Supply	3.3 V	Area	$16 \times 70 \times 80\,\mu m^2$
Current	$16 \times 18\,\mu A$	Capacitive load	$16 \times 2 \times 100$ fF
LF gain	82.5–83.6 dB	Phase margin	70°–60°
GBW	48–74 MHz	SR	68–120 V/μs
FoM_1	533–822 MHz pF/mA	FoM_2	755–1333 V pF/(μs mA)

$FoM_1 = GBW \cdot C_{load}/I$, $FoM_2 = SR \cdot C_{load}/I$

small-signal operation of M3 and M4. In the presented design, a value of K = 2 has been chosen as a good compromise between performance enhancement and stability. Table 6.3 summarizes the simulated performance of the implemented FD RFC OTA.

To integrate the dynamic scaling of noise technique into the proposed reconfigurable I-ΔΣ ADC, certain unit OTAs are required to dynamically power down during a single conversion and then power up again before the start of the next A/D conversion. In order to enable fast start-up, the biasing network of the OTA is not turned off and the cascode transistors M0b, M11 and M12 are used as switches to power down the core circuitry of the associated unit OTA.

Common Mode Feedback Circuit

A tradeoff applies to the design of the SC CMFB circuit as the CMFB capacitors C_{CMFB} sizes are comparable with those of the unit integrator. Therefore, using a CMFB for each unit element would significantly impact the OTA's driving capabilities, slew rate and GBW. To avoid this problem, a single adapted CMFB circuit, shown in Fig. 6.7, is shared among all unit integrators.

Here, the CMFB output drives the gates of all M9a and M10a transistors of the unit OTAs. By examining the operation of the CMFB circuit, the steady-state value of the CMFB output turns out to be independent on the parasitic capacitance at the output node (depicted as C_P in Fig. 6.7) [10]. However, as C_P increases, the start-up

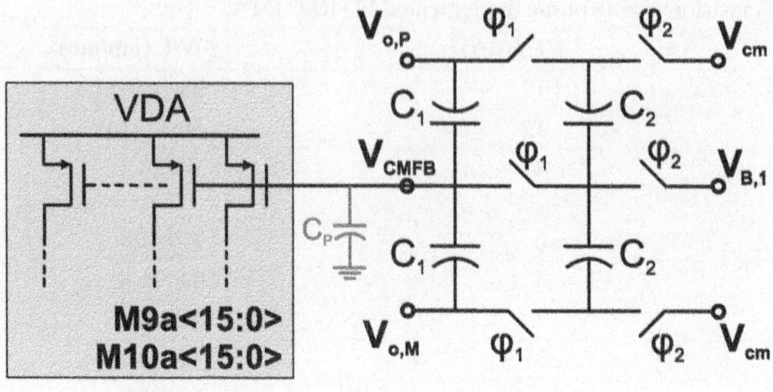

Fig. 6.7 SC common-mode feedback circuit shared among the unit OTAs of the scalable first integrator. A parasitic capacitance C_P arise due to the loading of the multiple current sources M9a and M10a. This capacitance does not influence the steady-state value of the CMFB output, but only lowers the start-up time

settling time of the CMFB circuit increases. In the implemented oversampling ADC, the start-up time is much lower than the duration of a single A/D conversion in the operational mode with the fastest conversion rate. Besides, increasing the ratio of the CMFB capacitors (i.e. C_4/C_3 in Fig. 6.7), results in lower start-up time and lower charge injection and leakage errors [10]. In this work, $C_3 = C_4/2 = 106\,\text{fF}$ is chosen.

6.2.2 *Shared Bootstrap Circuit for Input Switches*

Highly linear bootstrapped input switches are required for the utilized SC integrator implementation. However, using two conventional single-ended bootstrap switches for each integrator unit would be power and area inefficient. An alternative would be to share a single switch among all the integrators but this would increase the parasitic capacitance at this shared node. Therefore, in the presented design, a single bootstrap circuit, adapted from [11], is used, in which an array of NMOS switches (M0[15:0] W/L = 2.5 μm/0.35 μm, cf. Fig. 6.8) shares the same gate and input signals but the output signals are connected to their corresponding unit integrator.

It is important to mention that during Φ_1, the capacitor C_{bat} acts as a single battery across the shared gates and sources of M0[15:0]. Therefore, C_{bat} is scaled considering such increased loading effect and other parasitic capacitance that results from circuit interconnects. Consequently, a good compromise between performance, circuit area and required driving strength is achieved. By separately clock gating the signals applied to the gates of M0[15:0], individual switches can be accessed in order to be disable when the corresponding unit integrator is not needed. Consequently, the loading on the ADC driver decreases towards lower resolution modes.

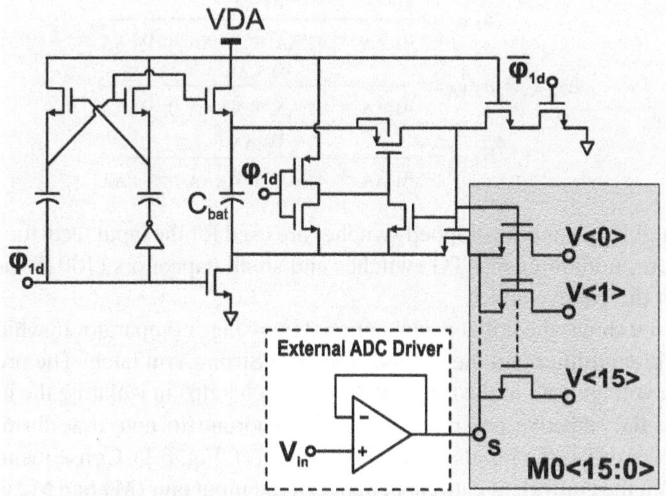

Fig. 6.8 Shared bootstrap circuit for the input switches in which an array of NMOS switches M0[15:0] shares the same gate and input signals but the output signals V[15:0] are connected to their corresponding unit integrator

For the remaining switches in the first integrator, as well as other switches in the CMFB circuit, second and third integrators, standard cells of TG switches are utilized. The TG switches are designed with low on-resistance that ensures the dominance of the OTA's g_m in defining the integrators time constant. Besides, the TGs sizing is done while considering the worst case corners due to the expected influence of the piezoresistive effect on the TG's on-resistance.

6.2.3 Voltage Summer and Comparator Circuits

In this work, a passive SC adder is utilized as a voltage summer for the input signal, as well as the output signals of the first, second and third integrators. Unlike an active adder (AA), the passive adder undergoes charge redistribution rather than actual voltage amplification. This mandates the translation of the modulator coefficients (cf. b_4, a_1, a_2 and a_3 Fig. 5.7) to match the transfer function of the passive adder (PA) and enable proper capacitor sizing. This is done using the following normalization relations:

$$a_{1,PA} = \frac{a_{1,AA}}{a_{1,AA} + a_{2,AA} + a_{3,AA} + b_{4,AA}}, \tag{6.11}$$
$$a_{2,PA} = a_{3,PA} = \frac{a_{2,3,AA}}{a_{1,AA} + a_{2,AA} + a_{3,AA} + b_{4,AA}},$$
$$b_{4,PA} = \frac{b_{4,AA}}{a_{1,AA} + a_{2,AA} + a_{3,AA} + b_{4,AA}}.$$

Besides, highly linear bootstrapped switches are used for the input feed-forward path. Furthermore, minimum size TG switches and small capacitors (100 fF) are used to implement the passive adder.

Figure 6.9 shows the utilized conventional two-stage comparator in which the first stage is a preamplifier and the second stage is a StrongArm latch. The preamplifier provides a voltage gain in the range of 8–12, which helps in isolating the high-speed latch from the sensitive passive adder. It is important to note that during Φ_1, the passive adder drives the inputs of the comparator (cf. Fig. 6.1). Consequently, charge sharing with the equivalent gate capacitance of the input pair (M1 and M2 in Fig. 6.9) occurs.

In this context, several measures have been taken to improve the accuracy of the passive adder. First, minimum size devices are chosen for the input pair of the preamplifier in order to reduce charge sharing. Second, during Φ_2, the preamplifier stage is configured as a unity gain buffer where the gate and drain of the input pair and the top plates of the passive adder are shorted. Hence, input auto-zeroing is performed. More importantly, the load capacitance of the passive adder (i.e. the gate-drain capacitance of the input pair) is cleared thus, the next summation starts with minimum signal dependent charge residue. Third, the StrongArm latch is triggered after the regeneration time of the preamplifier and near the end of Φ_1 to avoid any disturbance from propagating to the passive adder during the high-speed latching process. Fourth, the bottom plates of the passive adder capacitors are resetted during Φ_2 to the supply voltage, ensuring that the input pair being NMOS transistors start their operating region excursion at the beginning of Φ_1 from the same operating

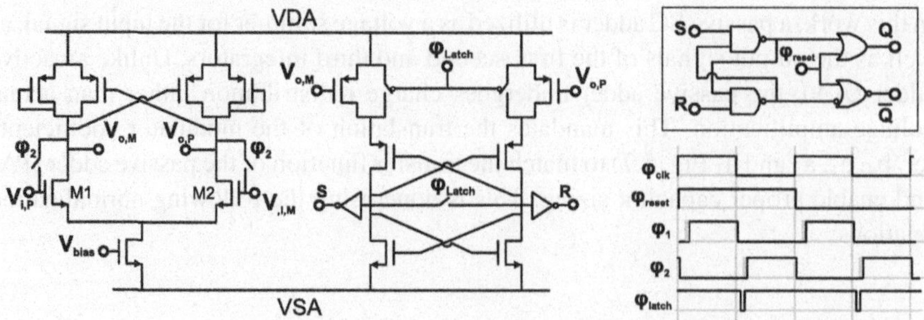

Fig. 6.9 Two-stage comparator. The first stage is a preamplifier while the second stage is a StrongArm latch. A conventional S-R latch follows the StrongArm latch and is power up using the digital supply voltage. The timing diagram of the control signals is shown on the right

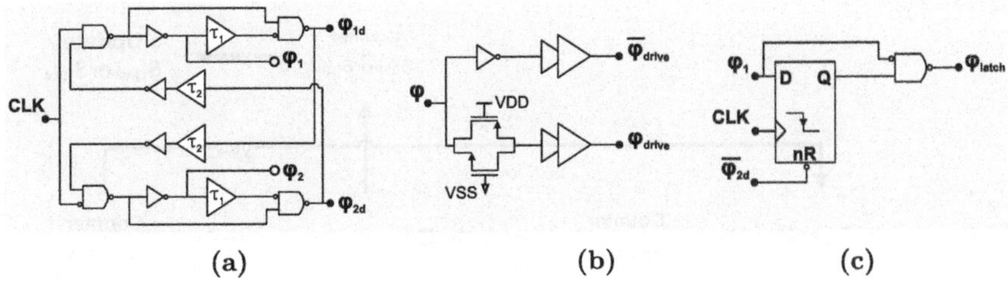

Fig. 6.10 **a** Non-overlapping clock signals generator in which τ_1 and τ_2 are used to adjust the non-overlap and delay duration [3]. **b** Alignment and buffering of the complementary clock signals. **c** Edge detector for controlling the StrongArm latch shown in Fig. 6.9

region (i.e. weak inversion). In another words, the common mode voltage of the passive adder is shifted to match the output common mode of the preamplifier.

6.2.4 Non-Overlapping Clocking Scheme

Non-overlapping clocking is utilized in the analog circuitry of the implemented I-$\Delta\Sigma$ ADC. This clocking scheme provides proper charge distribution in SC circuits [1]. Furthermore, bottom-plate sampling is used to ensure signal-independent charge injection. Figure 6.10a shows the clock generation circuit, which is used to set up the non-overlapping clock signals (Φ_1 and Φ_2) from an external clock source [3]. As shown in Fig. 6.10b, dummy TG resistors are used to align the complementary clock signals. Due to the distribution of the unit elements of the first integrator across the chip area, strong buffers are utilized for all the generated clock signals. For a higher-speed system employing similar unit integrators, distributed buffers and repeaters are necessary to meet the timing constraints. Figure 6.10c shows an edge detection circuit, which is used to generate the control signal for the StrongArm latch of the comparator at the end of Φ_1 (cf. Fig. 6.9).

6.3 Digital Control and Physical Design

6.3.1 Finite State Machine

The ADC chip includes a Finite-State Machine (FSM) for configuring the resolution modes and analog timing using a standard SPI. The number of reset (R cycles) and A/D conversion (M cycles) cycles are defined by two 8-bit registers. Additionally, each unit integrator has two corresponding 8-bit registers (S_{OFF}, S_{ON}), which control the on/off state of each integrator. When the unit integrator is in the on-state, a configuration for the onset of the dynamic scaling of noise is possible. Furthermore,

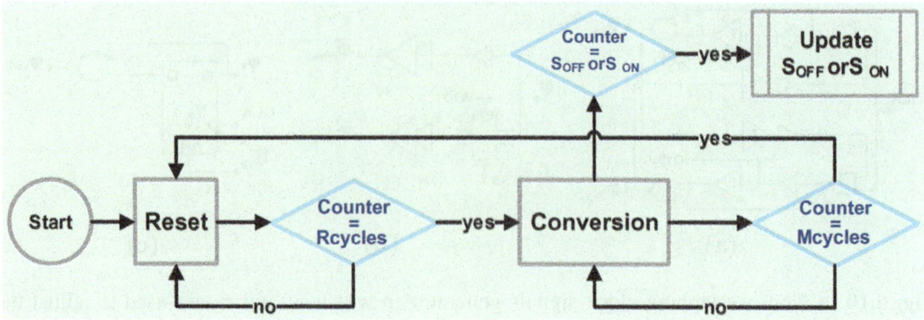

Fig. 6.11 Flowchart of the implemented finite-state machine (FSM). A 8-bit synchronous counter keeps track of the number of clock cycles in the reset (R cycles) and conversion (M cycles) states. Each unit integrator has two corresponding registers (S_{OFF}, S_{ON}) controlling the on/off state of each integrator. These registers are either initialized or dynamically updated during the reset and the conversion states, respectively

it is possible for the unit integrator to require few start-up cycles for properly turning on again. Therefore, this is done at the end of the A/D conversion as the last samples already contain large quantization errors.

Figure 6.11 shows a flowchart for the implemented FSM, in which two main states namely, reset and conversion, control the operational state of the I-$\Delta\Sigma$ modulator. A 8-bit synchronous counter keeps track of the number of clock cycles in the reset and conversion states. In the reset state, the I-$\Delta\Sigma$ modulator is resetted for a certain number of clock cycles (R cycles) and the on/off state of all unit integrators are initialized. In the conversion state, the A/D conversion takes place for a certain number of clock cycles (M cycles). Meanwhile, the FSM dynamically updates the on/off state of all unit integrators.

6.3.2 Floor-Planning and Layout

The distributed nature of the unit elements of the first integrator in addition to the expected nonuniform stress distribution in the ADC's UTC, require careful physical design and floor-planning of the ADC chip. For instance, by placing the circuit components far from the chip edges, large stress gradients are avoided [5]. In addition, the direction of the current flow in all analog and digital transistors is aligned to the same crystalline orientation [110] of a conventional (001) silicon wafer. Consequently, local uniformity and device matching are preserved in the presence of highly nonuniform stressors. Besides, the utilization of MIM capacitors, known for their stress-insensitivity [12], ensures the stress-independent realization of the $\Delta\Sigma$ modulator coefficients.

To address the piezoresistive effect and stress-induced errors on the performance of the digital circuitry, the timing constraints, such as the setup and hold times, are relaxed during the digital synthesis and place-and-route flow [12]. Besides,

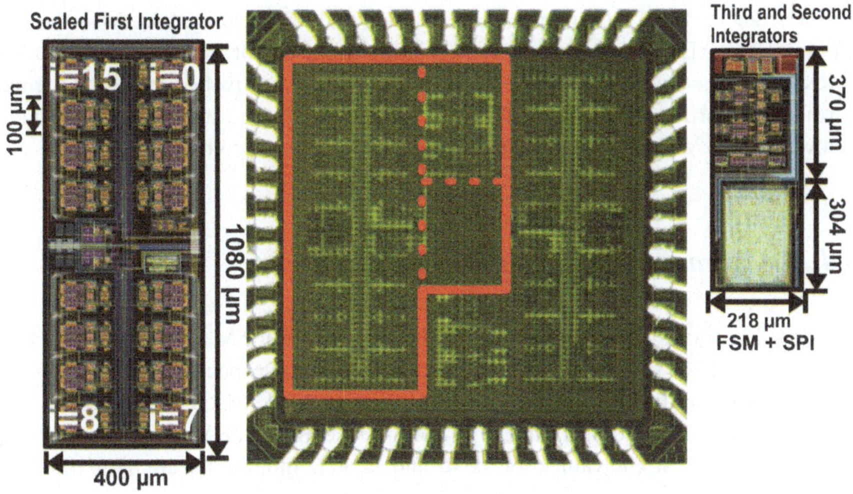

Fig. 6.12 Layout and die images of the reconfigurable I-ΔΣ ADC

the warpage of the free-standing UTCs is influenced by the passivation and metal interconnection layers, which exhibit negative and positive residual stress, respectively [13]. Therefore, special attention has been given to the routing density of each metal layer, ensuring homogeneous distribution of the digital signals and dummies in the different metal layers.

Figure 6.12 shows the layout and top-view image of the implemented reconfigurable I-ΔΣ ADC. The ADC occupies an area of about 0.5 mm^2 from which the digital controller consumes an area of $220 \times 300\,\mu m^2$. Each unit cell of the scalable first integrator occupies an area of about $200 \times 100\,\mu m^2$.

6.4 Measurement Results

6.4.1 Measurement Setup

After the ADC bare dies are received from the IC foundry, the chips are assembled onto a custom PCB shield. Conventional aluminum wire-bonding (25-μm thick diameter) is used to connect the ADC chip to the shield, which is mounted on top of an evaluation board.

For the ADC characterization, a fully integrated test instrument (ATX7006) for data converters is used. The digital I/O module generates the master clock signal and captures the ADC output bitstream, which is filtered using an off-chip CoI digital filter. The 20-bit arbitrary waveform generator (AWG20) and 16-bit dual reference generator (DRS16) slots have been used to generate the sinusoidal-wave input (2.28 Vpp) and reference (0 V and 3.3 V) signals, respectively. In addition, a first order buffered RC filter is used to filter wide-band noise and drive the FD inputs of the ADC. Separate supply regulators are used for the digital and analog cores, which have nominal voltage of 3.3 V and 3 V, respectively. Since high second harmonic distortion are observed during the measurements, the analog supply voltage is lowered to 3 V in order to optimize the operation of the prototype.

6.4.2 *Characterization of Bulk ADC Chips*

Figure 6.13a shows the measured ADC output spectrum for M = 133 and $S_{ON} = 3$. For $S_{ON} = 6$, Fig. 6.13b, c show the measured ADC output spectrum for the nominal 15b resolution for two different input frequencies of 4 kHz (deep inband signal) and 12 kHz (close to the Nyquist frequency). For a −3.2 dBFS (2.28 Vpp) input signal, the measured SNDR/SFDR values for the 4- and 12-kHz input tones are 88.5/104.2 dB

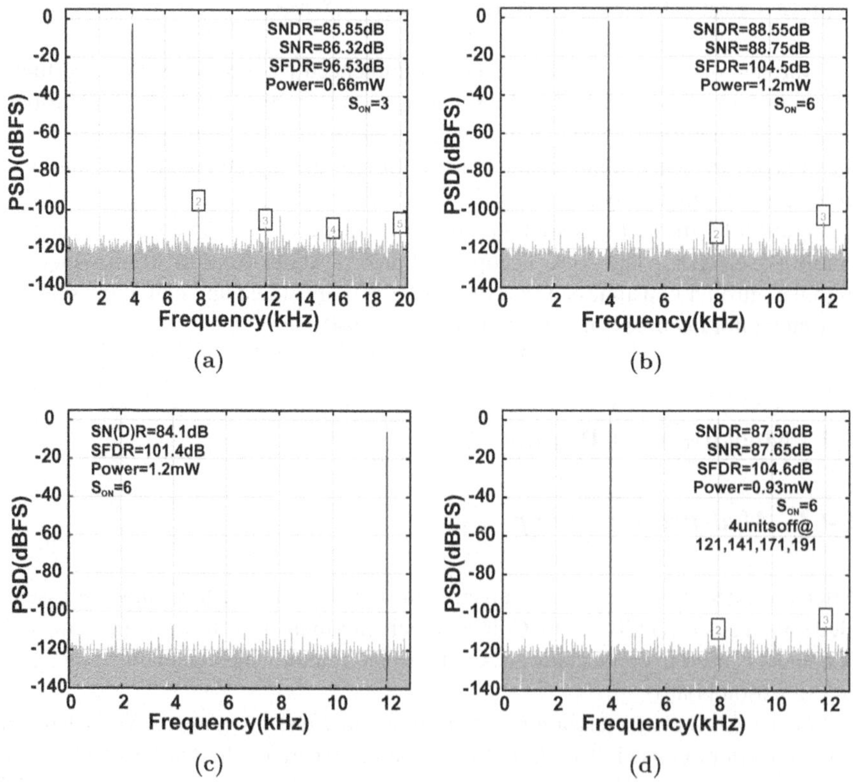

Fig. 6.13 Measured ADC output spectrum for the nominal 15 b mode of the proposed resolution reconfiguration. The ADC spectrum is for an input amplitude of −3.2 dBFS (2.28 Vpp) and input frequencies of 4 and 12 kHz. **a** M = 133 and $S_{ON} = 3$, **b–d** M = 211 and $S_{ON} = 6$. In **d** the proposed DR presetting is combined with the dynamic scaling of noise

and 84/101.4 dB, respectively. The absence of low frequency noise indicates proper integrator resetting.

Figure 6.13d shows the measured ADC output spectrum for the 15 b mode when the dynamic scaling of noise method is applied to the ADC. Here, 4 out of 6 unit integrators are dynamically disconnected within a single A/D conversion (as suggested in [14]), which illustrates the possible power savings (from 1.2 to 0.93 mW) at a minimum SN(D)R degradation.

For the DR presetting method, Fig. 6.14 shows the DR plots for different resolution modes at an identical input frequency of 4 kHz and a sampling frequency of $f_S = 5.4$ MHz. The ADC achieves a variable DR between 73 and 92 dB for the nominal resolution settings between 12 b and 15 b. Table 6.4 lists the DR presetting configurations used for each measurement. The measured DR points are chosen in increments of 0.5 b nominal resolution to demonstrate the ADC's flexibility and its DR fine selection granularity. Due to thermal noise, the 16-bit resolution mode is not measured. Thermal noise is confirmed to be the error source, and not quantization

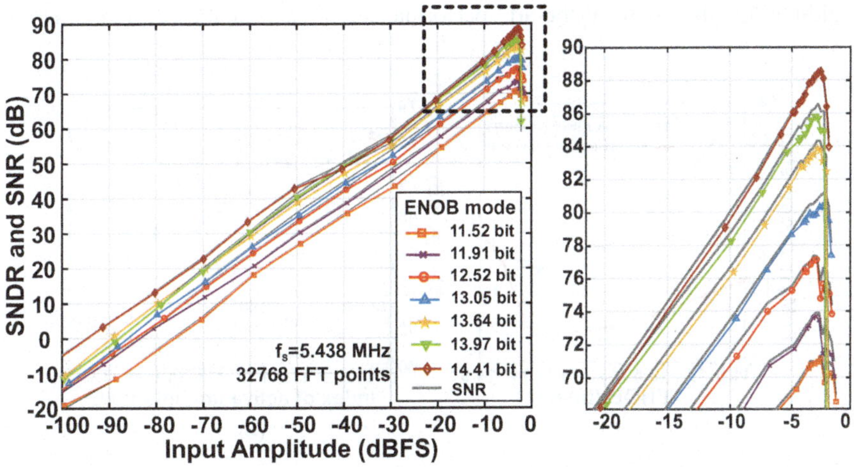

Fig. 6.14 Measured dynamic range of the proposed ADC for different resolution modes

Table 6.4 Summary of the configurations used for each ADC measurement

Resolution	S_{ON}	ENOB (bits)	M	f_N (kS/s)
15	6	14.41	211	25.8
15 + dynamic scaling of noise	6	14.24	211	25.8
15	3	13.97	133	40.8
14	2	13.64	110	49.4
14	2	13.05	83	65.5
13	1	12.52	77	70.7
13	1	11.91	63	86.3
12	1	11.52	54	100.7

or linearity errors, by averaging N measurements and observing an improvement of approximately 10 log(N) in the measured SN(D)R.

Figure 6.15a illustrates the measured FoM_W degradation when a simple reduction of the OSR is used at a fixed sampling frequency of $f_S = 5.4$ MHz. On the contrary, when the DR presetting is used (i.e. number of active integrator units is optimally chosen to the target resolution), a 2.5× improvement in the FoM_W is achieved.

Figure 6.15b plots the SN(D)R of the ADC when operated using sampling frequencies of $f_S = 5.5$ MHz and 6 MHz and OSR of M = 63. Here, a single unit integrator is activated for each data point. Hence, the performance of each unit integrator is evaluated separately. The index of the current active unit integrator is shown on the x-axis of Fig. 6.15b and the corresponding physical location on the chip is highlighted in Fig. 6.12. The least performing unit integrators (i = 7, 8, 9) are those located at greater distance than others from crucial circuit blocks, such as the clock generator, beta multiplier and CMFB circuit. On average, the unit integrators located on the right side (i = 0−7) perform better than the left side (i = 8-15).

Figure 6.16 visually compares the presented design against the state-of-the-art, highlighting the state-of-the-art performance in a scalabe design, thereby

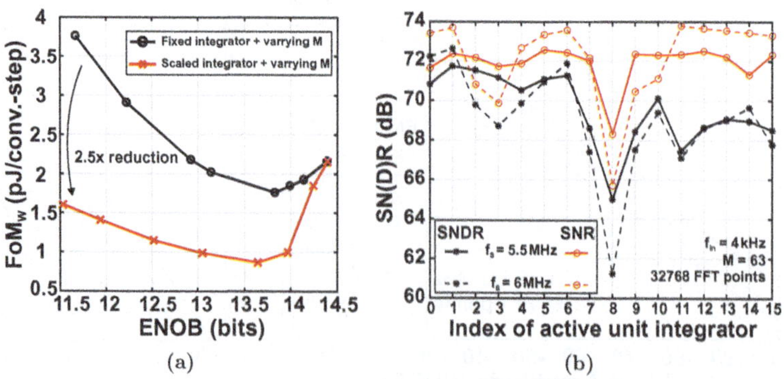

Fig. 6.15 **a** Measured effect of the first integrator scaling on the ADC FoM_W. Using optimum DR presetting, a 2.5× power reduction can be achieved in comparison to a pure OSR-based scaling of the DR. **b** Measured SN(D)R when the ADC is operated in the low resolution mode (M = 63) and a single unit integrator is activated for each data point. Consequently, the performance of each unit integrator is captured

Fig. 6.16 ADC performance summary and visual comparison with the state-of-the-art. (Data adopted from [15])

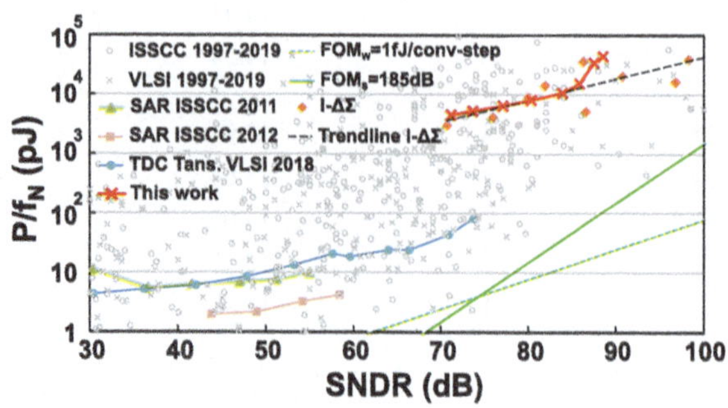

Table 6.5 ADC performance summary and comparison against the state-of-the-art

Ref.	This work						[14]	[16]	[17]	[18]	[19]	[20]
Architecture	I-ΔΣ						I-ΔΣ	I-ΔΣ	I-ΔΣ	I-ΔΣ	SAR	SAR
Reconfigurability	++						+	|	|	|	++	++
Technology (nm)	180						180	160	160	160	90	65
$Active area (mm^2)$	0.5						0.363	0.16	0.365	0.45	0.047	0.212
Supply (V)	3						3	1.8	1.8	1	1.1	0.4–1
Resolution (bit)	12	13	14	15	15+[14]	15	15	17	20	14	7–10	5–10
Power (μ W)	480	480	550	660	930	1200	1098	1650	6.3	20	8.2–17.4	0.23–0.21
Nyquist rate (kS/s)	100.7	70.6	49.4	40.8	25.8	25.8	200	40	0.0025	1.3	4000	5–2000
OSR	54	77	110	133	211	211	150	282	2000	1	1	1
DR (dB)	73	82	87	84	91	92	91.5	107.5	–	–	–	–
SNDR (dB)	71.2	77.1	83.9	85.8	87.5	88.5	86.6	98.3	119.8	81.9	56	30.4–55
SNR (dB)	71.3	77.4	84.4	86.3	87.6	88.8	88.2	104.4	119.8	81.9	–	–
SFDR (dB)	89.2	90.6	97.0	96.5	104.6	104.2	101.3	–	–	–	61	44–68
FoM_W $(fJ/step)$ *	1615	1160	880	1000	1860	2130	314	614	315	1480	16.4–6.5	143–22
FoM_S (dB) ** / ***	151.4/–	155.8/–	160.3/–	160.7/–	158.9/–	158.9/–	166.2/171.1	169.1/178.3	182.7/–	157.1/–	–/–	–/–

* $FoM_W = power/(2^{ENOB} \cdot 2 \cdot BW)$

** $FoM_S = SN(D)R + 10\log(BW/power)$

*** $FoM_S = DR + 10\log(BW/Power)$

demonstrating the efficiency of the proposed power scaling technique for I-$\Delta\Sigma$ ADCs. Table 6.5 summarizes the performance of the proposed ADC and compares it against the state-of-the-art in I-$\Delta\Sigma$ and reconfigurable ADCs.

6.4.3 Characterization of Ultra-Thin ADC Chips

Post-processing is performed on a number of bare die chips, which are back-thinned to achieve chip thicknesses of 100 and 30 μm. Figure 6.17 shows side-view images of the (from left to right) 700, 100 and 30-μm thick ADC chips. Since thin chips are known to experience higher stress levels compared to bulk chips, during packaging in particular, the thinned chips are carefully handled and assembled onto a flat custom PCB.

For comparison, Fig. 6.18 plots the SN(D)R of 4 dies from each thickness category, namely 700, 100 and 30-μm thick ADC chips. The same measurement conditions and ADC DR settings are applied to all chips, where input amplitude and frequency of −3.2 dBFS (2.28 Vpp) and 4 kHz, sampling frequency $f_S = 5.44$ MHz, $M = 133$ and $S_{ON} = 3$ are used. From the SNR plot, no deterministic signs of stress- or thinning-dependencies can be detected. However, the minimum SNDR achieved

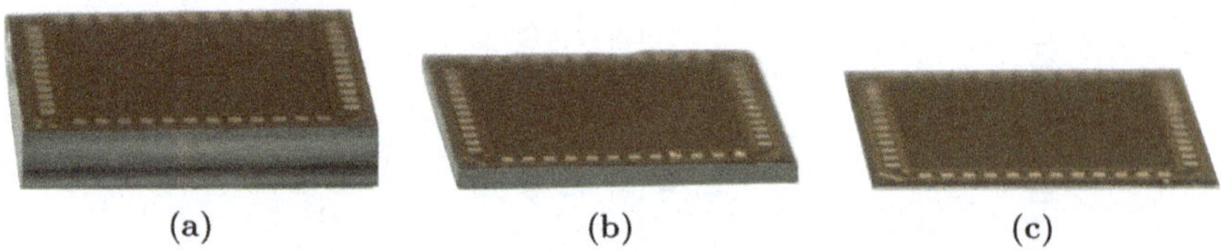

Fig. 6.17 Side-view images of the (from left to right) 700, 100 and 30-μm thick ADC chips

Fig. 6.18 Measured SN(D)R of 4 dies from each thickness category, namely 700, 100 and 30-μm thick ADC chips

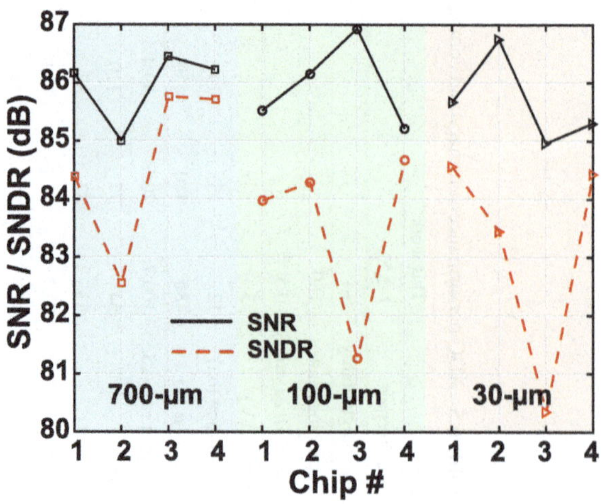

Table 6.6 ADC comparison against the state-of-the-art ADCs implemented on ultra-thin substrates

Technology	Single-crystalline Si UTC				Poly-Si	OTFT			InGaZnO	
Substrate Package	CFP, foil-compatible	foil Flex-IC	PI UTCP-foil	LCP foil	glass	foil-compatible	foil			
Ref.	This work	[21]	[22]	[12, 23]	[24]	[25]	[26]	[27]	[28]	[29]
Architecture	3rd I-$\Delta\Sigma$	SAR	SAR	SAR	2nd $\Delta\Sigma$	SAR	1st $\Delta\Sigma$	VCO	ADSM	SAR
Reconfigurability	++	|	|	|	+	|	+	+	+	|
Technology (μm)	0.18	0.18	–	0.5	3	20	–	5	15	30
$Active area (mm^2)$	0.5	–	–	–	26	700	260	19.4	27.9	27.5
Supply (V)	3	2.5	2.2	5	11.2	3	15	20	20	15
Resolution (bit)	12–15	8	12	10	10.6	6	3.8	7.7	6–8	5.7
Power (mW)	0.48–1.2	–	1.43	0.726	63.3	0.004	1.5	0.048	2	0.073
Nyquist rate (kS/s)	100.7–25.8	100	300**	250	400	0.01	0.016	167μ	0.3–0.01	0.027
OSR	54–211	1	1	1	128	1	16	1	–	1
SNDR (dB)	71.2–88.5	–	68**	59	65.6	–	-	–	40–50	35.9
SNR (dB)	71.3–88.8	–	–	–	–	–	26.5	48	–	–
SFDR (dB)	89.2–104.6	–	–	–	69	–	24.5	–	–	–
FoM_W $(nJ/step)$*	0.001	–	0.002**	0.004	14	–	3450	690	39–390	26

* $FoM_W = power/(2^{ENOB} \cdot 2 \cdot BW)$
** calculated from the product's datasheet

for each thickness category decreases with the chip thickness. Since the number of the measured ADC chips are limited, no solid conclusion can be drawn from this observation. Nonetheless, this SNDR degradation could point us towards the stress-induced variations in the on-resistance of the utilized switches, which results in higher nonlinearities.

Table 6.6 compares the performance of the proposed ADC against the state-of-the-art in flexible ADCs implemented using various TFT technologies, such as UTCs, OTFTs and InGaZnO inorganic TFTs. The presented reconfigurable I-$\Delta\Sigma$ ADC achieves the best power efficiency and highest resolution, which makes it a good candidate for HySiF integration targeting highly performing ultra-thin and flexible electronic systems.

References

1. Unbehauen R, Cichocki A (1989) MOS switched-capacitor and continuous-time integrated circuits and systems. Springer, Berlin, Heidelberg
2. Maréchal S (2014) Ultra-high resolution A/D converters: analysis, modeling and optimization tools, Ph.D. dissertation, École Polytechnique Fédérale de Lausanne (EPFL)
3. Understanding delta-sigma data converters, 2nd edn. Wiley-IEEE Press (2017)
4. Vogelmann P, Wagner J, Haas M, Ortmanns M (2019) A dynamic power reduction technique for incremental $\Delta\Sigma$ modulators. IEEE J Solid-State Circ 54(5):1455–1467
5. Mahsereci Y (2018, October) An ultra-thin CMOS sensor for in-plane stress sensing. Ph.D. dissertation, University of Stuttgart
6. Burghartz JN (2011) Ultra-thin chip technology and applications. Springer, New York
7. Wacker N (2013) Characterization and compact modeling of complementary metal-oxide-semiconductor transistors for flexible electronics. Ph.D. dissertation, University of Stuttgart
8. Hassan M-U (2017) Ultra-thin chip embedding and interconnect technology for system-in-foil applications. Ph.D. dissertation, University of Stuttgart, Shaker Verlag
9. Assaad RS, Silva-Martinez J (2009) The recycling folded cascode: a general enhancement of the folded cascode amplifier. IEEE J Solid State Circ 44(9):2535–2542
10. Choksi O, Carley LR (2003) Analysis of switched-capacitor common-mode feedback circuit. IEEE Trans Circ Syst II: Anal Digital Signal Process 50(12):906–917
11. Abo AM, Gray PR (1999) A 1.5-V, 10-bit, 14.3-MS/s CMOS pipeline analog-to-digital converter. IEEE J Solid State Circ 34(5):599–606
12. Mahsereci Y, Saller S, Richter H, Burghartz JN (2016) An ultra-thin flexible CMOS stress sensor demonstrated on an adaptive robotic gripper. IEEE J Solid State Circ 51(1):273–280
13. Angelopoulos EA, Al-Shahed MS, Appel W, Endler S, Ferwana S, Harendt C, Hassan M, Rempp H, Zimmermann M, Burghartz JN (2012) Manufacturing aspects of an ultra-thin chip technology. In 2012 Proceedings of the European solid-state device research conference (ESSDERC)
14. Vogelmann P, Haas M, Ortmanns M (2018, February) A 1.1mW 200kS/s incremental $\Delta\Sigma$ adc with a DR of 91.5dB using integrator slicing for dynamic power reduction. In: 2018 IEEE international solid-state circuits conference—(ISSCC), pp 236–238
15. Murmann B (2019) ADC performance survey 1997–2019. [Online]. Available http://www.stanford.edu/~murmann/adcsurvey.html
16. Gönen B, Sebastiano F, van Veldhoven R, Makinwa KAA (2016, January) A 1.65 mW 0.16 mm^2 dynamic zoom-ADC with 107.5 dB DR in 20 kHz BW. In: 2016 IEEE international solid-state circuits conference (ISSCC)

17. Chae Y, Souri K, Makinwa KAA (2013, February) A 6.3μW 20b incremental zoom-ADC with 6ppm INL and 1μV offset. In: 2013 IEEE international solid-state circuits conference digest of technical papers, pp 276–277
18. Chen C, Tan Z, Pertijs MAP (2013, February) A 1V 14b self-timed zero-crossing-based incremental $\Delta\Sigma$ ADC. In: 2013 IEEE international solid-state circuits conference digest of technical papers, pp 274–275
19. Yip M, Chandrakasan AP (2011, February) A resolution-reconfigurable 5-to-10b 0.4-to-1V power scalable SAR ADC. In: 2011 IEEE international solid-state circuits conference, pp 190–192
20. Harpe P, Zhang Y, Dolmans G, Philips K, De Groot H (2012, February) A 7-to-10b 0-to-4MS/s flexible SAR ADC with 6.5-to-16fJ/conversion-step. In: 2012 IEEE international solid-state circuits conference, pp 472–474
21. (2020) American semiconductor homepage. [Online]. Available https://www.americansemi.com/flex-ics.html
22. Sterken T, Vanfleteren J, Torfs T, de Beeck MO, Bossuyt F, Van Hoof C (2011, August) Ultra-thin chip package (UTCP) and stretchable circuit technologies for wearable ECG system. In: 2011 annual international conference of the IEEE engineering in medicine and biology society, pp 6886–6889
23. Elsobky M, Mahsereci Y, Yu Z, Richter H, Burghartz JN, Keck J, Klauk H, Zschieschang U (2018) Ultra-thin smart electronic skin based on hybrid system-in-foil concept combining three flexible electronics technologies. Electron Lett 54(6):338–340
24. Lin W, Lin C, Liu S (2009, November) A CBSC second-order sigma-delta modulator in 3μm LTPS-TFT technology. In: 2009 IEEE Asian solid-state circuits conference, pp 133–136
25. Xiong W, Zschieschang U, Klauk H, Murmann B (2010, February) A 3V 6b successive-approximation ADC using complementary organic thin-film transistors on glass. In: 2010 IEEE international solid-state circuits conference—(ISSCC)
26. Marien H, Steyaert MSJ, van Veenendaal E, Heremans P (2011) A fully integrated $\Delta\Sigma$ adc in organic thin-film transistor technology on flexible plastic foil. IEEE J Solid State Circ 46(1):276–284
27. Raiteri D, Lieshout PV, Roermund AV, Cantatore E (2013, February) An organic VCO-based ADC for quasi-static signals achieving 1LSB INL at 6b resolution. In: 2013 IEEE international solid-state circuits conference digest of technical papers, pp 108–109
28. Garripoli C, van der Steen JPJ, Smits W, Gelinck GH, Van Roermund AHM, Cantatore E (2017, February) 15.3 an a-IGZO asynchronous delta-sigma modulator on foil achieving up to 43dB SNR and 40dB SNDR in 300Hz bandwidth. In: 2017 IEEE international solid-state circuits conference (ISSCC), pp 260–261
29. Papadopoulos NP, De Roose F, van der Steen JPJ, Smits ECP, Ameys M, Dehaene W, Genoe J, Myny K (2018) Toward temperature tracking with unipolar metal-oxide thin-film SAR C-2C ADC on plastic. IEEE J Solid-State Circ 53(8):2263–2272

Chapter 7
Conclusion and Outlook

7.1 Conclusion

The main aim of this thesis is to introduce a design methodology for HySiF-compatible sensors and circuits. This aim has been demonstrated by implementing several ultra-thin passive and active electronic components. More specifically, this work is enabled by focusing on key factors, such as:

- Theoretical analysis and in-depth understanding of key building blocks (e.g. RH sensor, bandgap reference and ADC).
- Careful choice of CFP-compatible materials and fabrication process flow (e.g. PI for RH sensing).
- Prioritizing form factor, form flexibility and form adaptivity considering potential application for HySiF.
- Co-design of individual components by simultaneously considering the challenges on system, circuit and sensor levels (e.g. sense and reference capacitors for RH sensor, ADC topology optimally serving multiple sensors).
- Minimizing auxiliary components achieving complete independence for the implemented components.
- Pinpointing key performance metrics and suggesting several implementation of the same component to serve different applications.

The following part summarizes the main features and results from this thesis.

System-Level Concept—A generic model reflecting an envisioned HySiF concept has been introduced. Here, a centralized SoC UTC performs complex electronic tasks, meanwhile TFTs are distributed on the foil to preserve signal integrity and provide access to higher sensor nodes. Moreover, on-foil and on-chip sensors are complementary combined to customize the HySiF according to the targeted application.

Off-Chip/On-Foil Components—Various on-foil environmental and mechanical sensors, passive R, L and C components and TFT-based circuits are implemented on the surface of the CFP substrate. Such diversity of passive and active components is demonstrated on a flexible substrate that is used for UTC packaging.

M. Elsobky, *Ultra-Thin Sensors and Data Conversion Techniques for Hybrid System-in-Foil*, Recognizing Outstanding Ph.D. Research,
https://doi.org/10.1007/978-3-030-97726-9_7

For instance, an on-foil thin-film metal temperature sensor is fabricated on the BCB/PI substrate having nominal base resistance ranging from 80 Ω to 50 kΩ and temperature coefficients ranging from 2 to 3 mΩ/Ω °C depending on the substrate composition. Such on-foil temperature sensors, due to its low thermal mass, can be used for accurate temperature monitoring when the self-heating of the embedded UTC dominates or when the UTC package hinders the precision of the CMOS integrated temperature sensor.

Additionally, two variants of ultra-fast and wide sensitivity ultra-thin RH sensors are demonstrated. The electrochemical RH sensor, which is composed of thin-film nanosheets, showed no performance degradation when fabricated on the BCB/PI flexible substrate. This proves the suitability of the CFP to be used as a platform for the integration of a wide variety of sensors and not only polymeric sensors.

Furthermore, the capacitive RH sensor, which is composed of humidity-sensitive PI, is inherently compatible with the CFP process. It achieved a reliable operation and wide sensitivity (150%), which can be traded-off for ms response time, when needed.

Next, resistive strain gauges are printed using silver-ink on the BCB surface of the CFP substrate. Using four serpentine resistors, which are arranged in Wheatstone bridge configuration (surface area of $5 \times 5\,\text{mm}^2$), gauge resistances ranging from 400 Ω to 800 kΩ are achieved. The successful printing of such sensors shows the compatibility of the BCB/PI flexible substrate with recent metal printing techniques that are increasingly used in HySiF fabrication.

Moreover, on-foil planar spiral inductors with nominal inductance values ranging from 2 μH to 14 μH are designed to be compatible with the ISO/IEC 1443 class 1 and 3 standard for NFC. A peak Q factor of about 37 is achieved, which validates the fitness of the BCB/PI substrate in applications where MHz- and GHz-operations are needed.

For active components, a low-voltage OTFT digital library, including combinational and sequential logic, is characterized when fabricated on the BCB/PI substrate. The digital circuits operate using supply voltages in the range of 2 to 3.3 V. The shift register operates using a maximum clock frequency of 3 kHz. Besides, an on-foil sensor addressing circuit is developed utilizing the mentioned library, where biased-load inverters, shift register, pass transistors and analog switches are used.

Sensor Readout and Microcontroller UTCs—Ultra-thin smart sensor systems, sensor readout chips and microcontrollers have been successfully demonstrated. A humidity sensor system is implemented and its HySiF compatibility is tested by connecting the ultra-thin polymeric RH sensor to a 30-μm thick capacitive readout chip. A response-time of about 1 second is measured and no sign of stress-induced errors are detected for the 8-bit readout accuracy.

Furthermore, a strain gauge readout system that has been developed for robotic e-skin applications is characterized. It utilizes the CFP package as a HySiF platform in which UTCs, printed strain gauges and organic electronics are integrated. The signal amplification, offset-cancellation and digitization operations of the embedded UTC are tested.

Moreover, an ASIC prototype is implemented as a part of the MultiSense readout chip. The design and electrical characterization of a conventional bandgap reference circuit when integrated into a UTC are discussed. Here, stress-induced variations in the reverse-saturation current result in PTAT variations of the bandgap reference output voltage. For static stressors, a single point trim is sufficient. However, for dynamically varying stressors, this action is not enough. Therefore, inherently stress-insensitive components in combination with layout techniques should be used to design a stress-insensitive bandgap reference circuit.

Besides, an offset-compensated readout chip is designed where hardware resources are shared between the capacitive and voltage readout. The operation of the readout channel is tested using conventional 700-μm and thin 30-μm thick chips. No sign of thickness or stress dependency is detected in the analog readout operation.

Finally, the characterization of different blocks of a low power microcontroller upon thinning is discussed. For the accuracy of the on-chip 10-bit ADC, no sign of thickness or stress dependency is detected. However, the performance of the integrated temperature sensor slightly deviated from its nominal sensitivity values after the chip back thinning process. Additionally, the period jitter of the high-frequency RC oscillator significantly degraded. Note, that the stress-induced errors hint at the necessary re-calibration of ICs after the chip thinning process.

Reconfigurable ADC—After reviewing recently published reconfigurable, one concludes that SAR ADCs have dominated the low to medium resolution range. For ultra-thin or flexible ADCs, low-resolution SAR and $\Delta\Sigma$ ADCs have been widely implemented using various TFT technologies. Consequently, there is a need for a mechanically flexible reconfigurable ADC, which is tailored for medium to high precision multi-sensor systems. Therefore, third-order CIFF I-$\Delta\Sigma$ ADC architecture employing a 1-b quantizer and CoI digital filter is chosen. The DR presetting method is proposed for efficient hardware and resolution reconfiguration. The presented method is able to tune the ADC's DR through the scalable design of the first SC integrator. For an additional power-savings in high-resolution modes, the proposed ADC is able to use the dynamic scaling of noise technique. In this work, a simple solution is proposed for determining the optimal switching instances when using the dynamic scaling of noise technique.

In order to render the entire first integrator scalable, yet maintain its power-efficiency, various circuit blocks, such as the input switches' bootstrap circuits, are adapted. Furthermore, the OTA, sampling and integration capacitors and switches of the first integrator are divided into unit elements. Besides, simulation results showed the robust operation of the ADC against longitudinal in-plane stress in the ranges of $[-2.8, 1.6]$ GPa. From the measurement results, the FoM_W varies between 2.1 and 1.6 pJ/conv.-step over all nominal resolutions between 15 b and 12 b and a minimum FoM_W of 0.88 pJ/conv.-step for the 14 b mode is achieved. To this end, the performance of this ADC is comparable with state-of-the art fixed-resolution I-$\Delta\Sigma$ designs, while offering a scalable resolution range that cannot be covered by the normal SAR based reconfigurable ADCs. Interestingly, the proposed scaled integrator technique results are very good in mature CMOS technologies, which are still frequently used for sensor readouts, but also scales advantageously in DSM CMOS due to reduced parasitics in the circuit interconnects. Finally, back-grinding of bare die chips has been utilized to realize 100-μm and 30-μm chips. The measured SN(D)Rs of the thin ADC chips have shown no suggestion of thickness or stress dependency when the ADC chip is assembled on a flat substrate.

7.2 Future Outlook

HySiF is increasingly adopted in practical applications, including e-skin for prosthetics and robots and smart catheters and bandages. However, these applications represent niche markets, thereby call for more research and development in order to introduce the HySiF approach into mass production. To this end, improving the reliability, process yield and air-stability of on-foil components is of prime importance. Furthermore, the availability of models and process design kits (PDKs) aides the customization and wide adoption of the HySiF approach. In fact, more PDKs are being introduced for TFT technologies and foundries are promising a less than two-week manufacturing cycle. One expects the co-design of TFT and conventional CMOS circuits to enable optimal task and block allocation for each technology. By coupling the mechanical, thermal and electrically properties of the substrate, as well as, each HySiF component into a single software simulator, more mechanically compliant and electrical reliable HySiF can be achieved.

Besides, as more sensors are crammed into the foil, big data is generated and needs to be locally or remotely processed and communicated. Therefore, high-speed microprocessors and communication ICs are inevitably needed. Many questions arise regarding the economical feasibility of utilizing UTCs embedded in flexible packages or one can still utilize conventional rigid packages. By adapting the existing UTC embedding methods to fit into R2R manufacturing, high-speed UTCs can claim a central role in the mass production of flexible electronics.

Additionally, rethinking the design of basic sensors is needed so that less data is generated or even more meaningful signals are captured for less power and footprint. In this context, passive and active multi-modal sensors are good candidates for naturally encoding multiple sensor signals, thereby lowering the data overhead. The active multi-modal sensing variant can be achieved by employing split-gate electrodes and introducing the sensing material as the gate dielectric in TFTs. Consequently, signal amplification and buffering, in addition to local signal encoding, are achieved. In this context, deep-learning algorithms, implemented on TFT circuits or UTCs, turn out to be useful in order to decode the sensor signals and retrieve the corresponding physical variations.

In circuit design for flexible electronics, new challenges are faced, not only concerning the impact of mechanical stress on circuit blocks, but also due to other factors, such as parasitic RLC and cross-talk, which increase as long thin-film wire interconnects and thin substrates are used. TFT amplifiers and repeaters can be used to ensure analog and digital signal integrity across large-area systems. Additionally, the sensor readout should be able to handle elevated offset in the sensor signal, which traces back to higher mismatch and resolution of simple printing technologies. Furthermore, shielding the HySiF polymeric foil against electromagnetic interference is another challenge. Employing fully-differential signaling directly reduce the effect of environmental and mechanical disturbances. Moreover, the high internal resistance of flexible batteries limits peak currents, which are needed in high-speed circuits. Therefore, the developing of thin-film supercapacitors can be particularly important for mitigating non-idealities of flexible batteries.

For analog circuit design, ADCs in particular, the energy consumption, speed and accuracy have favorably scaled to support battery-operated electronic products. Although the speed of the implemented ADC is suitable for sensor readout, higher speed ADCs are essential in order to cope with the rapidly increasing number of sensors in the same system. Emerging techniques, such as ring amplifiers, time-interleaved and hybrid ADCs, can be utilized to unlock higher speeds using mature or advanced DSM technologies. As more UTCs are integrated into flexible electronic applications, the modeling and incorporation of stress effects in PDKs is of paramount importance. The characterization of high-resolution ADCs during elevated levels of mechanical stress can provide critical feedback for design improvements. Finally, noise analysis of stressed devices and the investigation of noise propagation in thin chips can improve the performance of ultra-thin microsystems.

Curriculum Vitae

Mourad Elsobky received the B.Sc. degree (with Highest Honors) in Information Engineering and Technology (IET) from the German University in Cairo (GUC), Egypt, in 2014, and the M.Sc. degree (Very Good) in Communications Technology from Ulm University, Germany, in 2015. He is the recipient of the German Academic Exchange Service (DAAD) scholarship in 2014. He is pursuing the Ph.D. degree at the University of Stuttgart, Germany and has been a researcher with the sensor systems department at IMS CHIPS, Germany. His research focus includes the design of ultra-thin hybrid sensor systems and power efficient data converters targeting IoT and flexible electronics applications. Mourad received the Best Paper award at the IEEE International Conference on Flexible, Printable Sensors and Systems (IEEE FLEPS) in 2019. He has published over 25 scientific articles and peer-reviewed several conference and journal papers such as the IEEE Journal of Solid-State Circuits and Organic Electronics—Elsevier.

M. Elsobky, *Ultra-Thin Sensors and Data Conversion Techniques for Hybrid System-in-Foil*, Recognizing Outstanding Ph.D. Research,
https://doi.org/10.1007/978-3-030-97726-9

The manufacturer's authorised representative in the EU is Springer Nature Customer Service Centre GmbH, Europaplatz 3, 69115 Heidelberg, Germany. If you have any concerns regarding our products, please contact ProductSafety@springernature.com

Printed and bound by CPI Group (UK) Ltd, Croydon, CR0 4YY
30/09/2026
02225465-0013